IEE POWER AND ENERGY SERIES 38

Series Editors: Professor A. T. Johns
D. F. Warne

THE
ELECTRIC
CAR

Other volumes in this series:

THE ELECTRIC CAR

Development and future of battery, hybrid and fuel-cell cars

Michael H. Westbrook

The Institution of Electrical Engineers

Co-published by: The Institution of Electrical Engineers, London,
United Kingdom

and

Society of Automotive Engineers, 400 Commonwealth Drive,
Warrendale, PA 15096-0001, USA

The Institution of Electrical Engineers,
Michael Faraday House,
Six Hills Way, Stevenage,
Herts. SG1 2AY, United Kingdom

Reprinted 2005

British Library Cataloguing in Publication Data

Westbrook, M. H. (Michael Hereward), 1926–
The electric car –(IEE power series; no. 38)
1. Automobiles, Electric 2. Automobiles, Electric–History
I. Title II. Institution of Electrical Engineers
629.2'293

ISBN 0 85296 013 1

Typeset by Mackreth Media Services, Hemel Hempstead
Printed in the United Kingdom at the University Press, Cambridge

Contents

Preface

In more than 33 years in the automotive industry I have had the good fortune to be involved with many developments in electrical and electronic technology that are now commonplace in today's road vehicles. The one area where to the outside world very little progress has been made is that of electric vehicles.

Hailed at various times as the answer to the adverse effect of the internal combustion engine on urban pollution and the generation of greenhouse gases, and in spite of considerable work by the automotive industry since 1990 under the spur of the California zero emission mandate, electric vehicles and particularly electric cars have failed to raise much interest in the car-buying public. Their limitations of range and performance are well known and have not been helped by the unsuccessful launching of electric cars of very limited capability in recent years. Considerable work has, however, gone into electric car and battery development in the last ten years with the prospect of substantial improvements in range and performance in battery cars as well as in hybrids and those using fuel cells.

I have been thinking of writing this book for some time, but I felt that progress in the last few years had been sufficiently encouraging to make it possible to predict that by 2025 the pressures of escalating oil prices and increasing concerns about pollution would make electric propulsion both economically viable and attractive to the car buyer and that a book on the subject would be of interest to a wide range of people. Over the last two years, therefore, I have attempted to gather all the available information on current electric vehicle technology together and set it out in what I hope is an easily readable form. I have also made my own predictions of the electric cars that will be available by 2025 when I would expect at least 25 per cent of all cars on the road to be electric in some form.

I would like to thank my wife for her forbearance with my disappearance nearly every afternoon for the last two years to my office to work on this book. I would also like to thank John West, who very generously has taken time from his busy life as a consultant on electrical machines and expert on electric vehicle drive trains, to read the first draft of the book. He has made many very valuable suggestions and I have incorporated these into the final text. I have also taken on board and implemented in the final text most of the constructive suggestions made by my referees, in particular by Ken Lillie.

I hope you will feel that after reading the book you have a better appreciation of where the technology is now and where it will go in the future, and that you will be enthusiastic to try these new developments in what is an old technology, when sensibly priced electric cars eventually become available.

Mike Westbrook
Colchester, February 2001

Acknowledgements

It is inevitable when writing a book covering a subject of such wide-ranging interest as electric cars that it is both necessary and essential to use material that has already been published in various forms. In particular I have used the diagrams, tables and photographs listed below with the permission of the copyright holders. I must particularly thank Intertech Corporation of Portland, Maine, USA, for permission to use in Chapters 4, 5 and 6 original material written by me and first published in their 1995 Intertech Report on Hybrid Electric Propulsion Systems. I am also very grateful to the major automotive companies listed below, who have been generous with technical information and photographs of their electric vehicles. In case there are any photographs which are not correctly acknowledged I can only say that every effort has been made to trace the copyright holders concerned.

The press and other departments of the following major automotive companies have been most generous in providing photographs and technical details of both their current and earlier electric vehicles: BMW AG, Citroen, Daimler-Chrysler AG, Fiat Auto SpA, Ford Motor Co., General Motors, Honda, Mitsubishi, Nissan, Porsche AG, Renault SA, Toyota, Volkswagen AG. Where photographs from these companies have been used they are acknowledged individually in the captions.

Permission to publish material from other publications and companies has been obtained as shown below:

- **Chapter 2**
  Figures 2.1, 2.2 and 2.3 are in the public domain and are published with attribution to *Scientific American.*
  Figure 2.4 is published with the permission of the UK Institution of Mechanical Engineers Library.
  Figure 2.5 and 2.6 are published with the permission of the Porsche Historic Archives.
  Figure 2.7 is published with attribution to *The Horseless Age* with the permission of 'Automotive Industries' and of Dr Ernest H. Wakefield and is reproduced from his book *History of the Electric Automobile* (SAE 1994).

- **Chapter 3**
  Figure 3.9 is published with the permission of the UK Institution of Electrical Engineers from a paper by B. Hredzak *et al.* 'Control of a novel EV drive with reduced unsprung mass', IEE Proceedings B, Electric Power Applications 1998.

- **Chapter 5**
  Figure 5.1 is published with the permission of the Advanced Lead-Acid Battery Consortium/International Lead Zinc Research Organisation (ALABC/ILZRO), Research Triangle Park, NC 27709-2036 USA from an ALABC/ILZRO paper by P. T. Moseley 'Electric Vehicles'.
  Figure 5.2 is published with the permission of Bolder Technologies Corporation, Table Mountain Drive, Golden, Colorado 80403, USA.
  Figure 5.5 is published with the permission of GM Ovonic, Maplelawn Drive, Troy, Michigan 48084, USA.
  Figure 5.6 is published with the permission of the UK Institution of Electrical Engineers from a paper by C. Vincent, 'Lithium batteries', IEE Review, March 1999, p. 65.

- **Chapter 7**
  Figures 7.9 (a) and (b) are published with the permission of the UK Institution of Mechanical Engineers from a paper by W. Steffens and M. Westbrook 'A microprocessor based method of battery charge measurement', IMechE Paper C204/85 presented at the International Automotive Electronics Conference, Birmingham UK, October 1985.
  Figures 7.11, 7.12 and 7.13 are published with the permission of the UK Institution of Electrical Engineers from a paper by J. Hayes 'Battery charging systems for electric vehicles', presented at an IEE Colloquium 'Electric vehicles – a technology roadmap for the future', London, May 1998.
  Figures 7.14 (a) and (b) are published with the permission of the Kansai Electric Power Co. Kitaku, Osaka, Japan.

- **Chapter 10**
  Figure 10.4 is published with the permission of the UK Motor Industry Research Association (MIRA) from a paper by P. Newman and P. Milner 'Minimising environmental impact, maximising marketability, the HDT approach', presented at an Electric Vehicle Seminar at MIRA, April 1992.

- **Chapter 11**
  Figure 11.1 is published with the permission of Johnson Matthey, Technology Centre, Blount's Court, Sonning Common, Reading RG4 9NH UK.

- **Chapter 12**
  Table 12.2 is published with the permission of the UK Institution of Electrical Engineers from a paper by J. West 'Propulsion systems for hybrid electric vehicles', presented at an IEE Colloquium on Electrical

Machines Design for All-Electric and Hybrid-Electric Vehicles, London, October 28th 1999.

Glossary

AC – alternating current
AFC – alkaline electrolyte fuel cell
ALABC – Advanced Lead-Acid Battery Consortium
APU – auxiliary power unit
BJT – bipolar junction transistor
CARB – California Air Resources Board
CSIRO – Commonwealth Scientific and Industrial Research Organisation
CVT – continuously variable transmission
DC – direct current
DI – direct injection
DOD – depth of discharge
DOE – US Department of Energy
EDLC – electrochemical double layer capacitor
EPRI – Electric Power Research Institute
Euro NCAP – European New Car Assessment Programme
EZEV – equivalent zero-emission vehicle
FCEV – fuel cell electric vehicle
GTO – gate turn-off thyristor
HEV – hybrid electric vehicle
HFAC – high frequency alternating current
IC – integrated circuit
ICE – internal combustion engine
IGBT – insulated gate bipolar transistor
IIHS – US Insurance Institute for Highway Safety
kWh – kilowatt-hour
Li-Ion – lithium-ion
MCFC – molten carbonate electrolyte fuel cell
MCT – MOS-controlled thyristor
MEA – membrane electrode assembly
MIT – Massachusetts Institute of Technology
MW – megawatt
$NaNiCl_2$ – sodium nickel chloride
NGP – next generation powertrain

NHTSA – US National Highway Traffic Safety Administration
NiCd – nickel cadmium
NiMH – nickel metal hydride
NOx – oxides of nitrogen
NPN – negative-positive-negative (transistor type)
PAFC – phosphoric acid electrolyte fuel cell
PEM – proton exchange membrane
PLI – plastic lithium ion
PM – permanent magnet
PNP – positive-negative-positive (transistor type)
PSOC – partial state of charge
PWM – pulse width modulation
PZEV – partial zero emission vehicle
RMS – root mean square
SCR – silicon-controlled rectifier
SOA – safe operating area
SOC – state of charge
SOFC – solid oxide electrolyte fuel cell
SOHC – single overhead camshaft
SR – switched reluctance
SULEV – super ultra-low emission vehicle
SUV – sport utility vehicle
TPV – thermophotovoltaics
ULEV – ultra-low emission vehicle
USABC – US Advanced Battery Consortium
VRLA – valve-regulated lead-acid
WHM – watt-hour meter
ZEV – zero-emission vehicle

Chapter 1

Introduction

This book has been written to provide a comprehensive picture of the history, current status and the likely future development of electric cars, as it is seen at the beginning of the second millennium. I have chosen to write about electric cars rather than electric trucks and buses as it seems to me that the success of electric cars in the market place will be the clearest signal of the technology reaching maturity. The category of cars includes such vehicles as taxis and light delivery vehicles but not trucks and buses.

The book is intended both for those working on all aspects of electric vehicle development and use, as well as for those members of the general public who have an interest in the technology and the future of personal transportation. It should also be of interest to those concerned by the pollution currently caused by the internal combustion engine and who see the electric car as one of the ways by which this pollution can be reduced.

In this book I have attempted to cover electric car development from a world perspective, although inevitably I have drawn mostly on my experiences in the automotive industry in Europe and the USA. I have used metric values for numerical data with equivalent imperial values where this seems appropriate. Where costs are shown they are in US dollars since they are easily translatable into local currencies across the world. In the vexed case of what units to use for fuel consumption I have used the metric measurement of kilometres/litre with the inverse European measurement of litres/ 100 kilometres also shown together with conversion factors for miles/ US gallon and miles/UK gallon.

The major automotive companies have been generous with information on their electric vehicle developments and I am indebted to them for both technical data and photographs of their vehicles as I am to the other companies who have sent data and photographs of their components. Full acknowledgement of this material is given under 'Acknowledgements' and against the item concerned where it appears in the book.

The book starts in Chapter 2 with the history of electric vehicle development. Here, the original scientific discoveries that made electric vehicles possible are described, together with the subsequent developments, which by the early years of the twentieth century had made what was initially a novelty into a popular practical road vehicle. Also described is the subsequent decline of electric vehicles when faced by the unstoppable rise of the internal combustion engine, and their renaissance in recent years when their low emission capabilities came to be recognised.

As is explained in Chapter 2, interest in electric vehicles reappeared in the 1960s, but the real trigger that started serious development of electric cars by the major automotive manufacturers was the sales mandate issued in 1990 by the California Air Resources Board (CARB). This required that by 1998, 2 per cent of light duty vehicle sales of each automotive company selling more than 35 000 vehicles per year in California, must be zero-emission vehicles (ZEVs). The required percentage then increased to 5 per cent in 2001 and to 10 per cent in 2003 with a lower sales threshold of 3 000 vehicles. The only practical way of meeting this zero-emission requirement was seen to be by the use of electric propulsion.

The consequence of these requirements passing into Californian law was to force the major automotive manufacturers to upgrade the relatively low-level activity then existing in electric vehicle development. The mandate required General Motors, for example, to sell about 35 000 electric cars in California by 2003. The response of the vehicle manufacturers was to launch specially developed electric vehicles that were intended to at least show their willingness to try to meet the original sales mandate. The first serious electric vehicle (EV) to become available was the General Motors EV1, originally known as the 'Impact'. This car with its lead-acid batteries had a maximum range of 145 km (90 miles) and is described in detail in Chapter 9. It was first seen at the Los Angeles Auto Show in January 1990 and was marketed in the USA from the autumn of 1996. Ford took a different approach and in 1991 announced an electric version of the European Escort light delivery van known as 'Ecostar'. This used a high temperature sodium-sulphur battery to give a range of 160 km (100 miles). A further alternative was proposed by a Swedish company called Clean Air Transport for whom International Automotive Design had developed an internal combustion/electric hybrid car known as the LA301. This vehicle had a range of up to 95 km (60 miles) on electric drive only and was intended to meet Californian zero-emission requirements over commuting distances.

The General Motors EV1 car was seriously marketed but only about 900 were sold or leased to the public between 1996 and January 2000. At this date production was stopped and the leased cars were withdrawn from the lessees. In the case of Ford a demonstration programme was started in 1993 with 103 Ecostar electric vehicles and a number of these vehicles are still in fleet service across the USA, Canada and Europe. Over 1.6 million km (1 million miles) have been accumulated since 1993 but there have been some problems with the sodium-sulphur battery overheating. The LA 301 was never offered for public use.

All this points to a less than enthusiastic promotion of electric vehicles by the major manufacturers and an indifferent reception by the majority of the general public. As a result of this and clear evidence that advanced batteries at an acceptable cost would not be available for some years, in 1996 CARB decided to drop their sales mandate for 1998 and 2001 and modify the original requirement of 10 per cent ZEVs in 2003 to 4 per cent plus a further 6 per cent made up either with more ZEVs or by using credits from partial zero-emission vehicles (PZEVs). These PZEVs would receive credits depending on their fuel economy and capability of operating as ZEVs over significant distances.

Since then a number of new electric and hybrid cars using advanced batteries have gone into low-level production and have become available for lease or purchase by the general public. These cars are listed and described in Chapters 9 and 10. Unfortunately they are being offered to the public at prices which are generally about twice that of the equivalent conventional car. Two exceptions are the Honda Insight and Toyota Prius hybrids, where heavy subsidies by the manufacturers have been used to make their prices competitive in the USA and Europe. All the indications are that significant reductions in costs are essential if electric vehicles are to have any chance of interesting the public and therefore becoming popular enough for the CARB 2003 mandate to have any hope of being met.

One of the objectives of this book is to describe the current status of electric vehicle development and to consider future developments that might lead within the next 20 years to a low-cost electric car. What is required is a car that will be attractive to the general public as an economical, effective and convenient solution to the pollution problem in urban areas. To achieve this low cost it is essential to reduce the cost of the propulsion motor and the battery or other energy source that provides energy to drive it. Those electric motors that are suitable for electric vehicle use are described in detail in Chapter 3 and an analysis of their advantages and disadvantages is made. The availability of low cost, lightweight electric motors suitable for vehicle propulsion depends largely on the development of low-cost, high-strength, permanent magnet materials and effective cooling methods that make it possible to use motors that are small and highly rated. Magnetic materials need to be suitable for use in permanent magnet (probably 'brushless DC') motors which can be combined with low-cost fixed gearing to enable the motors to operate at optimum speed and torque. Ultimately it would be highly desirable if it were possible to mount these geared motors in the drive wheels of the vehicle.

Obtaining efficient operation of the vehicle propulsion motors and coordinating this with the effective operation of both pure electric and hybrid vehicles requires sophisticated electronic controls that can be adapted to a wide range of operating conditions while at the same time optimising the efficiency and economy of what may be a very complex system. In particular, motor control and regenerative braking is entirely dependent on the electronic controls and the power electronics operating together as an integrated system. In Chapter 4 I have described these areas of the electric vehicle and how the control systems operate.

In the case of the energy source a great deal of money and a great amount of time have been spent on many attempts over the years to develop a low-cost battery with high-energy storage capability. The battery technologies resulting from this work are described in Chapter 5 and it will be seen that there is still a long way to go to compete with the gasoline tank as an effective energy storage device. There are, however, signs that the lithium-ion battery could be developed into a viable low-cost high-energy storage device since it offers not only high-energy density and a high cell voltage, but also appears to be easily adaptable to mass production methods. A detailed listing of the important data on all the battery types considered are shown in Table 5.1 but I must make clear that the data on large-scale production costs are my best estimates and not published figures.

Other energy sources that can store or generate electrical power in an electric vehicle, including fuel-cells and flywheels, are considered in Chapter 6. They offer some interesting possibilities particularly in the case of fuel-cells, but it is difficult to see how, in the short term, their costs can be reduced sufficiently to make them realistic contenders for widespread use.

The availability of suitable charging facilities both at home and in places where electric cars may be parked is not a trivial matter and may determine how effectively electric cars can be used by the public. I have discussed the issues in detail in Chapter 7 and made recommendations of the minimum requirements for a viable charging infrastructure. The charging problem is overcome if fuel-cells are used as the electric vehicle power source since then it is only necessary to store hydrogen or hydrocarbon fuel on the vehicle to feed the fuel-cell and there is no requirement for external charging. Hybrid electric vehicles also bypass the charging problem by carrying their own internal charger operated from their heat engine, albeit at a significant cost penalty.

The way in which electric vehicles are designed to meet the special requirements they have for maximum efficiency and safety is described in Chapter 8. Efficiency is particularly important because of the relatively small amount of energy that can be stored in a battery compared to that stored in a gasoline tank. This high efficiency is obtained by minimising weight, reducing rolling resistance by the use of high-pressure tyres and designing the vehicle body for minimum air resistance. Safety is particularly related not only to crash performance but also to the protection of the operator and service personnel from the high voltages (200–350 V) used in the battery, motor and control system.

Over recent years many battery electric cars have been designed using the technologies discussed in Chapters 2 to 8. The result is seen in Chapter 9 in which 11 production battery electric cars currently or recently available for hire or purchase from the major automotive manufacturers by the general public are listed, with four of these being described in considerable detail. Also listed are 16 prototype and experimental cars.

One approach to overcoming the range problem is the hybrid electric vehicle and this, with its many variants, is described in Chapter 10. As is shown there is very little agreement on the optimum arrangement in the case of the most widely used heat engine/electric hybrid, particularly in respect of the way in which the main and auxiliary power sources are combined. The objective is generally to minimise fuel consumption, but this may be modified by the need to provide a certain minimum range when only electric power is used to meet zero-emission requirements. The major problem with hybrid electric vehicles is the cost of having two propulsion systems and it is difficult to see how this can be overcome.

Although the fuel-cell itself has been described in some detail in Chapter 6, the various methods by which the hydrogen fuel it requires can be supplied and the infrastructure needed to do this is described in detail in Chapter 11. The many problems remaining to be overcome if we are to ever have a 'hydrogen economy' with hydrogen replacing gasoline and diesel as the fuel for our cars in the future is also discussed.

In Chapter 12 I have looked at the costs of complete electric vehicles and their components in an attempt to understand where these costs are likely to go by 2025 and therefore which electric vehicles are likely to be acceptable to the public by that date. The costs quoted are either those published by manufacturers or are my best estimates based on the general information available in publications, at conferences and on the Internet.

In Chapter 13, the final chapter, I have considered what are likely to be the developments over the next 20 years in electric vehicles, their components and the necessary infrastructure. If the increasing pressure for an improved environment is to be met and the rocketing cost of hydrocarbon fuels taken into account, it will be essential that viable low-cost electric vehicles are available. At the end of the chapter I have speculated on the types and features of the production electric cars that will be available to the general public by 2025 and given short descriptions of the four types of electric car I expect to fulfil that role.

In spite of the many setbacks that have happened in the development of electric vehicles up to the present, I believe that the signs are encouraging. It is therefore worthwhile to set down in detail what appear to be the technical, operational, infrastructure and financial issues related to the development of a viable electric car industry in the first quarter of the twenty-first century. I also hope that this book will help to promote an interest and enthusiasm for electric cars in the large number of people who follow the latest developments in cars and take an interest in the fortunes of the automotive industry. In my opinion, by the middle of the century electric cars will be the only effective answer for the majority of personal transportation in what will be an era of both high sensitivity to pollution and highly priced fuel.

Chapter 2

The history of electric cars up to 1990

The history of electric cars is the story of how the development of practical methods of storing electrical energy combined with the invention of methods of converting electrical to mechanical energy provided the possibility of a new, quiet and clean method of propulsion. This history also describes how that method of propulsion experienced a brief period of ascendancy at the beginning of the twentieth century, before being overtaken by the internal combustion engine.

The reasons for the reappearance of electric cars in the last two decades of the twentieth century are also discussed and give a lead into the subject matter in the remainder of this book.

2.1 The early days

The history of electric vehicles has to start with the first successful attempt to store electrical energy made by Alessandro Volta in Italy in 1800. Volta had been fascinated by the experiments of Luigi Galvani, Professor of Medicine in Bologna, who had observed that a frog's leg would twitch if he touched a muscle with a copper probe and the associated nerve with a zinc probe.

To simulate this condition Volta assembled plates of copper and zinc, separated by pasteboard soaked in salt water, and discovered that a continuous electric current could be drawn. By connecting a number of these copper/zinc cells together in series he found that he could obtain a larger electrical potential at a continuous current, or by placing the cells in parallel a larger current at the same potential. The primary battery had been born and was then called the Volta pile by subsequent researchers.

The next event of major significance was in 1821 when Michael Faraday demonstrated that a wire rod carrying an electric current supplied by a Volta pile would rotate around a fixed magnet if one end was unconstrained by allowing it to hang in a bath of mercury. He also showed that the magnet would rotate around the wire if the fixed and moving elements were reversed. Furthermore, the direction of rotation was reversed if the polarity of the electric current was reversed. The principle of the electric motor had been demonstrated.

It is said that at one of the famous Faraday lectures at the Royal Institution, Faraday demonstrated this phenomenon to an attentive audience of the great and good of the day. After the demonstration he was asked by the then British Prime Minister, the Earl of Liverpool: 'And Mr Faraday, of what use is it?' Faraday is said to have replied, 'Sir, its principle will someday form a useful commodity, and then you can be sure that the House of Commons will tax it'.

Another possibly apocryphal story is that Queen Victoria attended a similar demonstration, presumably with Prince Albert who was well known to have a great interest in scientific discovery, and asked a similar question on what use this discovery could possibly be. Faraday's reported reply on this occasion was 'Of what use is a baby?'.

Faraday went on to discover electromagnetic induction in 1831, so demonstrating the intimate relationship between magnetism and electric current and laying the foundation for electrical and electronic technology including the electric motors and generators required for electric cars.

By 1832 an operating electric motor had been demonstrated in Paris, and in 1835 a small operating motor was demonstrated in London by Francis Watkins. This motor operated by having a bar magnet mounted on a shaft rotating inside stationary coils of wire in which the electric current was successively switched by contacts on the shaft. This produced a synchronised rotating magnetic field – the principle still used in many of today's electric motors. Further developments in the following year demonstrated that this motor could be used in useful ways for drilling and driving a lathe.

All these developments led inventors to think of the possibility of using the Volta pile to drive a motor attached to the driving wheels of a light-weight vehicle; this is reported to have been achieved in 1835 when a small model car was built and demonstrated by Professor Stratingh in Groningen, Holland. There is also a report that between 1834 and 1836 Thomas Davenport in the USA built and demonstrated an electric road vehicle, and in 1837 a carriage built by R. Davidson moved under electric power in Aberdeen. However, a more reliable report is of the German physicist Moritz Jacobi operating a Volta-pile-powered paddle boat on the river Neva in St Petersburg in 1838.

Apart from these occasional demonstrations which showed that electrical propulsion was possible but limited in application, little further is known to have happened until 1859 when a seminal event occurred. This event was the invention of the lead plate/sulphuric acid/lead plate cell familiar to us today as the lead-acid starter battery used in all of today's family cars, and as a motive power battery in many of today's electric cars.

In the course of the development of electrochemistry between 1800 and 1859, it had been shown that if electrodes connected to a Volta pile were immersed in water, oxygen bubbles formed on the positive electrode and hydrogen bubbles formed on the negative electrode, and these gases could be collected in test tubes. In 1802 N. Gutherot had connected a galvanometer across the electrodes after such an experiment had been completed and detected a small current flow in the opposite direction to that of the originally applied current.

This observation does not appear to have been acted upon until 1859 when Gaston Planté, a Belgian experimenting in the field of electrochemistry, placed two layers of lead sheet separated by cloth into a container of dilute sulphuric acid and observed that an electric current could be drawn from this cell, and as this happened lead sulphate formed on the positive plate. If a current was then passed through the cell from a Volta pile, the lead sulphate would decompose, and after disconnection of the electrical supply further current could be drawn in this reversible process. This was the first effective secondary cell in which charge and discharge processes could be repeated many times.

The earlier experiments with electric motors in the 1830s were to lead nearly 30 years later to the 'ring' direct current motor invented by Antonio Pacinotti in Italy in 1861. Apart from the invention of the motor itself, he made the very important discovery that if the motor was rotated mechanically a current was generated in a reverse direction to the current applied when it was being run as a motor. The electromechanical generator had been born. This discovery would make possible a continuous supply of significant amounts of electrical energy.

This discovery was built on by a Belgian, Zenobe Theophile Gramme, and in 1869 he was able to construct the first direct current electric motor. This could provide mechanical power of more than one horsepower (746 W), and could be used as a generator if mechanical energy was supplied to rotate it. Gramme was also the first to connect a steam engine to drive one of his generators and so to provide significant amounts of electrical power. These developments were closely followed by the Siemens brothers, initially in Germany and then subsequently in England, where in 1870 they developed and patented the Double T Iron Armature motor/generator.

The availability of this new technology signalled the birth of the electrical supply industry. This started by supplying continuous electrical current for electroplating. The industry then expanded to provide electricity for electric-arc street lighting, and then for domestic and general lighting with the invention in 1879 of the filament lamp by Thomas Alva Edison in the USA and Joseph Swan in England. However, of more interest to us in the context of this book is the rapidly developing availability and capability of both storage batteries using the Planté cell and electric motors capable of providing sufficient mechanical power to propel a road vehicle.

2.2 The first road vehicles

In 1873 it had been shown by R. Davidson in Edinburgh that it was possible to drive a road vehicle, in this case a four-wheel truck, using an iron/zinc primary battery. However, it was not until 1881 in France that G. Trouvé made the first electric vehicle to be powered by a secondary Planté battery. The vehicle was a tricycle and used two modified Siemens motors which drove one large propelling wheel through two chains, these motors developed about 1/10 hp and propelled the 160 kg (350 lb) vehicle at about 12 kph (7 mph). Trouvé also operated an electrically powered motor boat on the Seine in the same year, showing the versatility of this new form of power.

In the following year of 1882 in England, Professors William Ayrton and John Perry demonstrated an electrically powered tricycle which used ten lead/acid (Planté type) cells in a battery of 1½ kWh capacity giving 20 V to a ½-horsepower direct current motor mounted beneath the driver's seat. The vehicle is shown in Figure 2.1. The vehicle speed was controlled by switching the batteries one after the other in series. The range was reported as between 16 and 40 km (10 and 25 miles) depending on the terrain, and the maximum

Figure 2.1   Ayrton and Perry tricycle of 1882 ('Scientific American')

speed as 14 km/h (9 mph). This was still three years before Carl Benz was to demonstrate the first operating internal combustion engined vehicle, also a tricycle. The vehicle had one further claim to fame: it was the first vehicle to have electric lighting. Two small filament lamps each of four candlepower and providing 50 lumens of light flux, were mounted to illuminate, respectively, an ammeter and a voltmeter and can be seen in Figure 2.1 near the top of the drawing. They were supplied from the traction batteries.

With its top speed of 14 km/h (9 mph), the Ayrton and Perry tricycle fell foul of the notorious 'Red Flag Act', otherwise known as The Locomotive (Roads) Act of 1865. This had been introduced in the United Kingdom to curb what was considered to be the excessive speed made possible by the use of steam propulsion on the roads, and the resulting disturbance caused to horses.

The terms of the Act are given below:

1. Home Office powers to control locomotives abolished.
2. At least three persons to be in charge of a locomotive, with an extra person if two or more wagons are drawn.
3. One person to precede the locomotive at a distance of at least 60 yards, on foot and showing a red flag as warning. This person also to assist horse traffic.
4. Locomotive to stop if the man in front, or anyone in charge of a horse, so signals.
5. Locomotive drivers to give as much room as possible to other traffic.
6. Whistling and letting off steam within sight of horses prohibited.
7. Two lights to be carried in front, one at each side.
8. Penalty for breaches of any of the above provisions, £10.
   Speed limits 4 mph in open country, 2 mph in towns and villages. Penalty for breach: £10 for driver, whether owner or not.
9. All locomotives to bear the name and address of the owner.
10. Local authorities of the City of London, the metropolis, boroughs and towns with more than 5000 population empowered to make special orders as to hours of operation and speeds, so long as the speed does not exceed 2 mph. Penalty for breach £10.
11. Existing Acts prohibiting the use of steam engines within 25 yards of a road not to extend to ploughing engines.
12. A man to be stationed in the road to warn of an engine in a field.
13. Notwithstanding all other provisions, all locomotives to be liable to prosecution if they make a nuisance. Nothing in the Act to affect the right of any person to recover damages for injury from the use of a locomotive.

This Act introduced very severe restrictions on all mechanically propelled vehicles on British roads and until its repeal in 1896 undoubtedly set back the development of electric vehicles in the United Kingdom compared to the progress beginning to be made in other countries. It did not, however, stop Gustave Phillipart having an electric tramcar built in Belgium in 1882 and shipping it to London for experimental trials. Whether this was the same 'electric bus' reported to be on the streets of London in 1886 is not clear, but evidently the pace of development was increasing as in 1887 Radcliffe Ward assembled an electric cab which ran in Brighton. History does not tell us if anyone was brave enough to hire it!

The USA now began to come into the picture again, when the first electric vehicle since that of Thomas Davenport in 1834–6 was built by Philip W. Pratt in 1888 in Boston. This was soon followed in 1890 by Andrew L. Riker of New York who demonstrated a 150 lb tricycle using a 1/6-horsepower motor and a 100 lb battery to give a range of 48 km (30 miles) at 13 km/h (8 mph). In 1895 Riker went on to build a four-wheel electric vehicle weighing 140 kg (310 lb). This used two ½-horsepower motors each driving a rear wheel. With a 60 kg (135 lb) battery this vehicle was capable of carrying one person at 19 km/h (12 mph) for 4 h.

There were other inventors and manufacturers such as Barrows and Holtzer-Cabot in the USA by this time, but perhaps the most significant were the partners Morris and Salom who in 1895 started to build their two-seater Electrobats. This led to the 1896 Electric Road Wagon shown in Figure 2.2 and very rapidly on to a series of coupes and hansoms for use as taxis in New York with protection against the weather for the passengers, although not the driver.

This taxi service was to start on November 1ˢᵗ 1896 with 12 hansoms and one coupé. The vehicles had front-wheel drive using two ½-horsepower motors and had the doubtful benefit of rear-wheel steering. Power was from 44 lead-acid cells giving 88 V and the range was claimed to be up to 48 km (30 miles). There appears to have been quite an enthusiastic reception given to this new form of transportation by the younger members of society, although concern was expressed about the exposure of passengers in a vehicle which did not have a horse in front to act as a partial shield from the common gaze. However, many people saw it as a fad which would soon pass and not to be compared to the thrill of travelling in a hansom with a frisky horse.

It was at this time that some of the descriptive terms used in the automotive field were coined, particularly 'horseless carriage' first used by E. P. Ingersoll in his USA publication *The Horseless Age* in November 1895, and 'automobile' first used in the *Pall Mall Gazette* of London in October 1895.

2.3 Competition for speed and reliability

By 1895 increasing interest had also developed in competitions between internal combustion-engined vehicles to show which was superior in speed and reliability. These were undertaken over increasing distances, culminating in Europe in the 120 km (75 mile) Paris–Rouen reliability trial of 1894 in

Figure 2.2   Morris and Salom electric road wagon of 1896 ('Scientific American')

which 13 steam and eight internal combustion engined vehicles took part, with one of each type finishing at an average speed of 19.9 km/h (12.36 mph).

In June 1895 the Paris–Bordeaux Race, which involved a return trip of 1 135 km (705 miles) without rest, was run. Because of the distance to be covered there was intense interest in the newspapers of the day and this was heightened by the fact that an electric vehicle driven by Charles Jeantaud was taking part for the first time. Although the electric vehicle did not win, and in fact only completed the first half of the race to Bordeaux, and this by dint of changing batteries at battery stations every 24 miles, its performance highlighted the competition between steam, gasoline and electricity in the race for success as the propulsive medium of the future.

Not to be outdone, *The Chicago Times-Herald* promoted an 85 km (53 mile) race from Chicago to Evanston. It is not certain how many vehicles took part, but when they set out on Thanksgiving Day, November 18th 1895 with six inches of snow on the road, the line up of vehicles included a Morris and Slalom Electrobat and a Sturges electric vehicle. A gasoline-powered vehicle was the winner, the electric vehicles completing about 32 km (20 miles), probably their realistic range without recharging or a battery change in the conditions, particularly bearing in mind the adverse effect of low temperatures on battery capacity.

The aristocracy were beginning to take an interest in the new methods of travel and in 1896 G. Julien, a Spanish engineer, designed an electric Victoria for the Queen of Spain, which was built by Thrupp & Maberley coachbuilders of London. The vehicle had front-wheel steering and used a primary battery weighing 90 kg (200 lb) to power the propulsion motor. It also had electric lighting using filament lamps, a very advanced feature for this date.

Although the use of a primary battery might be seen as a step backwards, the interesting feature of the battery was that it is reported to have used a zinc plate in an alkaline electrode and in use the zinc electrode was consumed. This sounds remarkably similar to the modern zinc-air battery described in Chapter 5. Claims were made that the battery could operate the carriage at 10 mph for 60 h, which seems unlikely with the conventional battery technology of the day, but with what must have been a relatively light vehicle, would be feasible with a modern zinc-air battery!

Another electric carriage using interchangeable dry batteries was built by Walter Bersey in London in the same year. It used two motors driving each of the rear wheels through a chain, seated two people and had a claimed range of 56 km (35 miles) at 19 km/h (12 mph). The battery interchange capability was particularly interesting, since this was the method used by Charles Jeantaud in the Paris–Bordeaux race of 1895, and was to be used quite widely over the following 15 years for 'touring' by electric car, particularly in the USA.

The French were proving to be particularly inventive in the electric car field, exemplified by the electric coupé demonstrated by M. A. Darracq in Paris in 1897. This vehicle was notable because it used regenerative braking for the first time. This braking method makes use of the property of the electric drive motor to act as a generator charging the battery when overdriven mechanically by the vehicle wheels. This recharging loads the drive motor/generator sufficiently to provide a powerful braking effect, and at the same time increases the energy stored in the battery enough to increase the vehicle range by up to 10 per cent. Some earlier vehicles had used electric motor braking but dissipated the energy in a resistor rather than feeding it back to the battery.

In 1897 the 'Red Flag Act' was repealed and electric vehicle design blossomed in the United Kingdom as it already had done in France and the USA. The introduction of electric taxis already made in the previous year in New York was extended to London, so that by the summer of 1897 there were 13 taxis plying for hire in New York and 15 in London. The use of electric taxis in congested city streets had much to recommend it, as indeed it could do today, for the emissions from the then most widely used power source, the horse, were different but equally unpleasant to those to be experienced in our time from the internal combustion engine.

The taxi application is, in fact, almost ideal for the use of electrical power, since it requires vehicles to travel short distances with frequent stops, and gives reasonable opportunities for overnight battery recharging. This resulted in a number of the electric vehicles built at this time being designed for use as taxis. An example is the Victoria Hansom Cab shown in Figure 2.3 and built by the American Electric Vehicle Company in 1898 under the direction of Clinton E. Woods of Chicago. This vehicle had two motors driving hard-rubber-tyred rear wheels, with enough battery capacity to give a range of 30 miles. It also had electric lights and a rudimentary form of heating within the enclosed passenger cabin consisting of electric foot-warmers for each passenger.

Figure 2.3   Victoria Hansom Cab built by the American Vehicle Company in 1898 ('Scientific American')

2.4 Electric vehicles compete with steam and gasoline

While all these developments in electric vehicles were going on the gasoline-fuelled internal combustion engined vehicles were also being rapidly developed, so that by 1900 the market for automobiles was almost equally divided between the three contenders of steam, electricity and gasoline. In that year in the USA, of the cars manufactured, 1 684 were steam-driven, 1 575 were electric, and 963 were propelled by gasoline engines.

Meanwhile, in 1899, the capability of electric vehicles had been highlighted by the capture of the world speed record by an electric vehicle. This vehicle, shown in Figure 2.4, had been built by Camille Jenatzy in France, and designed specifically for racing. It used a streamlined aluminium/tungsten alloy body, pressed steel wheels, and had the new pneumatic tyres which had been proved superior to solid rubber tyres by M. Michelin in the 1895 Paris–Bordeaux Race. In the summer of 1899 Jenatzy went for the record in front of a large Parisian crowd, achieving 98 km/h (61 mph) over the measured distance. This record was to stand for three years until beaten by a gasoline-engined vehicle.

Enthusiasm for the electric vehicle was strong at this time, primarily because of its quietness compared to the then unsilenced gasoline-powered vehicles, and its ease of starting without the tiresome need to hand crank the engine. This enthusiasm was sufficient for Thomas Edison to tackle the problem of finding a better electric vehicle battery, a quest that still goes on

Figure 2.4   Jenatzy electric racing car 'Jamais Content'. Holder of the world road speed record of 98 km/h (61 mph) from 1899 to 1902 (IMechE Library)

today. By 1901 he had invented the nickel-iron battery which is described in detail in Chapter 5. Unfortunately this battery, although an improvement on the lead-acid battery in energy stored per unit weight of about 40 per cent, suffered from the high cost of the materials used in its construction which precluded its use except in special applications. In more recent years the nickel-iron battery has been used fairly widely in stationary applications where long life and reliability are essential.

2.5 The golden age

The 12 years from 1900 to 1912 was the golden age for electric vehicles, although gasoline-powered vehicles were developing rapidly over this period. By 1903 there were more electric vehicles in London than those powered by the internal combustion engine, but this situation was not to last. In 1906 the Ford Model K appeared and constituted the first real challenge. By 1909 when the Model T was launched it was already clear that the race for the new market for personal transport had been won by the internal combustion engine. This did not stop the number of electric vehicles in use in the USA increasing to a peak of 30 000 in 1912. However, by this date there were 900 000 gasoline-engined vehicles on the road in that country, and comparable numbers in Europe. When the crank handle was first replaced by the self-starter in the same year, and the silencer was introduced to reduce the engine noise substantially, it was game, set and match to the internal combustion engine.

Also in this 1900–1912 period, ideas to improve range and performance continued to appear and these developments are exemplified in the 1900 French Electroautomobile and the 1903 Krieger electric-gasoline car, both of which used an electric motor and a gasoline engine in a hybrid configuration to provide a combined drive to the front wheels. The latter car also had a form of power steering.

Innovative manufacturers such as Ferdinand Porsche were also developing electric cars, in particular the 1900 Porsche No.1 Lohner-Wagen (see Figure 2.5) which used two electric motors in the wheel hubs. He also developed racing cars such as the 1902 Lohner-Porsche Rennwagen shown in Figure 2.6 that carried 1 800 kg of batteries to enable it to compete in long-distance races. With its enormous battery load and its four 1.5 kW wheel motors it must have been a formidable machine both to drive and to stop! It is of interest to note that the particular machine shown was bought by a Mr E. W. Hart from Luton, England and he is shown driving the car with Ferdinand Porsche beside him. At about this time Ferdinand Porsche was also responsible for the design of one of the first experimental electric/internal combustion hybrid vehicles. In this vehicle, called the 'Mixt-Wagen', an auxiliary gasoline engine drove a generator which charged the storage batteries which then in turn powered the front-wheel electric motors.

Figure 2.5   Porsche No.1 Lohner-Wagen of 1900 with wheel hub motors ('Porsche Historic Archives')

Figure 2.6   Lohner-Porsche racing car of 1902 ('Porsche Historic Archives')

Typical of the electric cars of 1912 is the Century Electric Roadster shown in Figure 2.7. This car shows the advances in design made in both electric and gasoline-powered vehicles over the preceding ten years. Instead of looking like adapted horsedrawn carriages, both electric- and gasoline-powered cars were beginning to develop a functional design suitable to the propulsion method and the requirements of the user. In the Century Electric Roadster the motor drives the differential directly and resistance control switching gives a relatively smooth control of speed at the cost of some loss of range. However, this ease of operation could not compensate for the much lower cost of the gasoline-powered cars of the time. The Century Electric Roadster was on sale in 1912 at $1 750 while the Model T could be bought for only $550.

At about this time of maximum production of electric cars, the combination of a gasoline engine with electric drive had been developed to a stage where a production car could be marketed. This was the Woods Gasoline-Electric of 1916. In this car, a small four-cylinder gasoline engine was directly coupled to an electric motor-generator which was then coupled to the rear drive axle by a conventional propeller shaft. With this arrangement, whenever either the motor-generator or the gasoline engine is operated both must rotate. This means that with suitable manual control switching the car may operate as a straight gasoline vehicle, as a straight electric vehicle, or as

Two Passenger
Century Roadster **$1750**

Figure 2.7   Century Electric Roadster of 1912 ('The Horseless Age')

a combined gasoline-electric vehicle with the electric motor drive supplementing the mechanical drive, or in the generator mode recharging the battery. This arrangement is what we now call a parallel hybrid vehicle, and it offers the possibility of unlimited range combined with quiet and pollution-free operation in the city.

2.6 Cost problems for electric drive

Unfortunately for the future of this type of vehicle the cost disadvantage versus the straight gasoline vehicle was even greater than for the pure electric. The Woods Gasoline-Electric cost $2 650, nearly $1 000 more than a straight electric vehicle and a massive $2 100 more than a Model T. Quiet and easy to operate it may have been, but this price difference made it inevitable that it would only sell in small numbers. It is interesting to note that this cost problem still exists with modern hybrid vehicles and this is discussed in more detail in Chapters 10 and 12.

From 1912 onwards the number of electric cars in use and being built would steadily decline, while the number of gasoline cars would increase dramatically. This trend was accelerated by the 1914–18 Great War in which the use of gasoline-powered vehicles at the front under very difficult conditions confirmed their reliability and utility.

The last new model electric car to be built in the USA before the modern era was assembled by the Automatic Transmission Company of Buffalo, New York in 1921. It was a two-seater called the Automatic, with a top speed of 25 mph and a claimed range of 60 miles. It sold for $1 200 when the Model T cost less than $300 and brought to an end a period of almost forty years in which many technical developments had been made. These technical developments would be resurrected in another 50 years, when the electric vehicle again became a possible way of overcoming some of the environmental problems associated with the gasoline engine.

2.7 The dark ages (1925–1960)

By the early 1920s almost all the electric car manufacturers had either gone out of business or started to make cars with gasoline engines. An exception was the Baker Electric Company of Cleveland, Ohio, which stopped building electric cars in 1921 and started to specialise in electric industrial trucks. The company still exists in the form of Baker Electric, a subsidiary of United Technologies, and makes, among other things, electric fork lift trucks.

By the mid-1920s the gasoline-engined vehicle was in total ascendancy all over the world, with 20 million a year being built in the USA and probably half that number in Europe. A small number of electric cars were still being made to special order into the 1930s but even this stopped by 1935. To all intents and purposes the electric vehicle industry was dead.

There were a few exceptions to this dark picture: in Japan, because of wartime restrictions on gasoline, electric vehicles were used between 1937 and 1954. In 1949 there were 3 299 electric cars in use; about 3 per cent of all Japanese vehicles on the roads at that time. In the United Kingdom in the late 1940s and early 1950s the use of electric vehicles for milk delivery was established and they are still used for this service today. Some 20 000 of these vehicles have been built and most are still in use. This particular delivery application is ideal for an electric vehicle, the requirement being for a quiet vehicle with a range of around 30 km (20 miles) and up to 200 stops, completed within a working day, and with controlled overnight battery charging.

2.8 The modern era

2.8.1 The 1960s

From 1960 onwards interest in electric cars began to appear again, air pollution caused by gasoline engines was beginning to be of concern and a number of small firms were set up to try and meet what was seen to be a new demand. Most vehicles produced were conversions of conventional mass-produced cars; for example, in 1960 the Eureka Williams Corporation in Illinois built 150 Henny Kilowatt electric cars which were conversions of the four-door Renault Dauphine. In the mid-1960s Scottish Aviation Ltd built three specially designed Scamp electric vehicles and major vehicle manufacturers such as GM and Ford undertook experimental conversions of gasoline vehicles to electric drive.

A serious attempt to produce a car suitable for urban use was made by the Enfield Company in England in 1966. The Electricity Council, with responsibility for the regional Electricity Boards and the Central Electricity Generating Board in England and Wales, had decided to evaluate electric car technology. The Enfield 8000 electric car, shown in Figure 2.8, was produced in 1966 by a small specialist company and was designed throughout as an electric car, rather than being an electric conversion of an existing gasoline-engined vehicle. The car chassis was built of steel tube with a square cross-section and motive power was provided by eight lead-acid traction batteries and one auxiliary battery. It had a 6 kW series-wound DC motor which was controlled by switching the batteries in series and parallel to give 12, 24 and 48 V. The motor field was in two sections which could be switched in series and parallel at each voltage to give smooth control of speed. No attempt was made to use regenerative braking.

The weight of the vehicle was 975 kg and it had a maximum speed of 64 km/h (40 mph) and a range of between 40 and 90 km (25 and 56 miles)

Figure 2.8   Enfield 8000 electric cars in 1966

depending on driving conditions. This performance was felt to be adequate for the urban use test programme proposed by the Electricity Council.

Sixty-six of these cars were purchased in the early 1970s and distributed to the 12 Electricity Boards in existence at that time in England and Wales, and also to the Central Electricity Generating Board and the Electricity Council. An analysis of the results of three years' use made at the beginning of 1979 produced some interesting statistics and highlighted a number of problems that occurred when inexperienced operators used the vehicles. The total fleet mileage over the three years was 404 787 km (251 577 miles), an average of just over 6 000 km (3 700 miles) per car. This average concealed large variations in distance travelled. Generally where a number of cars were operated from the same depot, they were well maintained and the battery regularly charged and relatively high distances were covered. When the car was the only one at a depot, battery maintenance tended to be neglected and only small distances were covered. If the car was used for commuting and the driver obtained a direct benefit, then the car and the charging were well looked after and enthusiasm for electric cars was evident. The average energy consumption of all the cars was 300 to 400 Wh/km (480 to 640 Wh/mile), somewhat higher than would be expected from an electric car of the 1990s of this size and weight. Although useful data was obtained from these long-term tests which continued until 1980, the cost of the vehicles, which was £2 000 when a basic BMC Mini cost £1 000, precluded the continuation of the tests and the sale of more cars to the general public. Before production stopped in 1976, 112 of the cars had been built.

The major automotive companies, which had been converting a few conventional gasoline-driven vehicles to electric drive, now began to look at the possibility of designing an electric car from scratch. A good example of this was the Ford Motor Company in Britain, then a relatively autonomous arm of Ford Motor Company in the USA. The Ford of Britain Research Staff were asked early in 1966 to look at the possibility of designing a small electric car for urban use with the following specific characteristics:

1. Small enough to occupy minimum road and parking space
2. High manoeuvrability
3. Minimum pollution
4. Simple to operate
5. Low initial and running costs

By June 1967 the experimental prototype was completed and given the name Comuta.

The completed car is shown in Figure 2.9, and was capable of seating two adults in the front and two children in the back, this being achieved within

Figure 2.9   Ford Comuta electric car of 1967

an overall length of 80 inches and a height of 56 inches. It was claimed that three Comutas could be parked side on in the same space as one large conventional car.

The Comuta was powered by two series-wound DC motors each driving a rear wheel. Each motor weighed 18 kg (39 lbs) and was 14 cm (5.5 inches) in diameter. A chopper system provided pulse control of the motors. The chopper system used thyristors controlled by a simple electronic logic unit; an early application of electronics to electric cars.

Power was obtained from four lead-acid batteries weighing a total of 170 kg (380 lbs) and providing 120 ampere-hours at a five-hour rate. This was claimed to give a range of 64 km (40 miles) at a steady 40 km/h (25 mph), maximum speed being quoted as 64 km/h (40 mph). A small amount of heating was provided in the cab by circulating the waste heat from the motors and control unit, but it was accepted that this was inadequate for use in winter.

When the prototype was unveiled in June 1967 tests were undertaken on range, acceleration, ride and handling, and noise performance. A second prototype was also built and tested by Ford personnel in the USA. Results showed that noise from the gear assembly between motors and wheels, and from the chopper control system was higher than was desirable, and modifications were made to overcome this. Some ride and handling problems caused by the small wheelbase were evident, although range and acceleration were as expected with the motor and battery combination used. Generally, although the car performed well within its expected limitations, it did not seem to offer enough to be worth further development. One of the prototypes still exists and is on display in the Science Museum in London.

In the USA at this time developments were also taking place in electric car technology. In the mid-1960s General Motors started a $15 million programme which resulted in prototypes called the Electrovair and the Electrovan. These vehicles used a newly developed three-phase AC drive system in which power from either a silver-zinc battery or a fuel-cell was converted to three-phase AC using electronically driven silicon-controlled rectifiers (SCRs). GM also converted an Opel Kadette to DC drive using both lead-acid and zinc-air batteries. The latter battery was claimed to give the car a range of 240 km (150 miles) at 48 km/h (30 mph).

In 1968 General Electric demonstrated the GE Delta which used nickel-iron batteries to give a range of 64 km (40 miles) and a top speed of 89 km/h (55 mph). In the same year Ford demonstrated an experimental E-car using a nickel-cadmium battery supplying a 52 hp DC motor. Apart from the very valuable development of AC drive, this period had only proved the difficulty of producing an electric car of acceptable range and performance at a cost users would be prepared to pay.

2.8.2   The 1970s

An attempt to overcome the cost problem was made by the Sebring-Vanguard Company in Florida when in 1972 they produced a low-cost two-seater called the Citicar (see Figure 2.10). The car had a range of only 25 miles and a top speed of 45 mph, but it was sold at the remarkably low price of $3 000, similar to a low specification gasoline car of the time. As proof of how critical cost is in making electric cars saleable, about 2 500 of these low-performance cars were sold before production stopped in 1976. This was the largest number of viable electric cars of the same design ever sold in the modern era until Peugeot Citroen exceeded this number in the late 1990s.

In the mid-1970s a number of other small companies in the USA were engaged in converting conventional cars to electric drive. These included Jet Industries, McKee Engineering and Electric Vehicle Associates (EVA). This last company offered an electric conversion of a Lancia with a range of 96 km (60 miles) and capable of speeds up to 96 km/h (60 mph). Over a period of seven years 185 cars were sold, but by 1982 the company had stopped production and closed. Some larger companies also felt that because of their special interests they should have a stake in this technology.

Figure 2.10   Sebring-Vanguard Citicar of 1976

This resulted in electric vehicles such as the Globe-Union Endura (1977); the GE Centennial (1978); and the GE/Chrysler ETV-1 (1979).

In the United Kingdom, in the same period of the mid-1970s the Electraction company was founded in Essex by ex-British Ford personnel to exploit what was seen (incorrectly as it turned out), as a niche market for specialist electric vehicles. Four fibre-glass-bodied vehicles were produced: the two-door Precinct; the open top Tropicana; a Beach Buggy; and an Electric Box Van. In spite of much publicity, the company failed to sell significant numbers of these vehicles and closed in 1979.

Although not a car, one of the most interesting developments was the conversion, by Lucas, of the Bedford van to electric drive. This conversion was fully developed for production and built on the same line as the conventionally powered Bedford van. About 300 vehicles were built, but production stopped when the £4 500 ($6 750) government subsidy being given for each vehicle was withdrawn.

In continental Europe, the 1970s were also an active period. In 1973 Electricité de France converted 80 conventional cars to electric drive, while in Germany Daimler-Benz and Volkswagen had built experimental vehicles. In Italy in 1975 Fiat had designed an experimental prototype: the X1/23B. This was a two-seater using lead-acid batteries and DC drive with a range of 48 km (30 miles) and a top speed below 64 km/h (40 mph).

In Japan, Daihatsu, Toyota, Mazda and Mitsubishi all worked on prototype electric vehicles during the 1970s, assisted by Japanese Government funding of $20 million for five years from 1971. Early vehicles used DC drives with lead-acid batteries and AC drive systems did not appear until the 1980s.

2.8.3 The 1980s

Governments became interested in the 1980s in the possible environmental advantages of electric vehicles, and this triggered the availability of official funding for electric vehicle programmes. This led to the Ford/GE programme sponsored by the US Department of Energy in the mid-1980s, which resulted in the development of the ETX-1 vehicle using an advanced AC drive system. Tubular lead-acid batteries connected to provide 200 V were used with a Darlington transistor-based inverter producing up to 300 A AC to drive a 37 kW two-pole induction motor.

By July 1988 Ford and GE had developed the AC drive system of ETX-1 into an advanced powertrain using the sodium-sulphur batteries then becoming available in useful quantities and originally invented in the Ford Research Laboratories, with a transistor inverter driving a synchronous permanent magnet motor. This motor, with an integral two-speed gearbox, was mounted concentrically on the rear axle of the vehicle, see Figure 2.11. The drive system was installed in two ETX-2 vehicles which with their sodium-sulphur batteries had a range of 160 km (100 miles) and were capable of

Figure 2.11   Ford integral electric motor gearbox transaxle of 1988

speeds up to 96 km/h (60 mph). These two vehicles were delivered to the US Department of Energy in December 1988.

In the United Kingdom, Clive Sinclair started to work on a low-cost electric car, encouraged by the UK Government's decision to exempt small electric vehicles from taxation. The resulting three-wheel vehicle, launched in January 1985, could hardly be called a car; it was a single-seater scooter with a 250 W motor, a relatively small lightweight lead-acid battery and a maximum speed of 24 km/h (15 mph). The driver was unprotected from both the elements and the other traffic, and from the highly publicised launch on there were major concerns about its safety in traffic. There also appeared to have been little thought given to battery performance under the frequent deep discharge conditions existing in electric vehicle use, with the result that performance and battery life were poor. About 10 000 units were built in its short production run, of which 6 000 were still left in stock when the production company closed in October 1985. Some C5s are still raced in special events, but the C5 cannot be considered to be a serious electric vehicle capable of comparable performance to a conventional car.

In France by 1988 there were about 500 electric vehicles in experimental use and in the hands of customers. Most of these were Peugeot 205 or Citroen C15 van conversions. In Germany the experimental vehicles of the late 1970s had been further developed and by 1988 GES had developed a VW Golf conversion called the City Stromer. The vehicle was designed to meet the European safety standards of the time, used a thyristor chopper DC drive with electronic control, and featured regenerative braking which was claimed to give a 5 per cent increase in the 90 km range. In the same period in Italy, Fiat undertook further development of electric cars which resulted in the Fiat Panda Elettra conversion, while in Japan by 1988 DC drive had been superseded by AC drive using both lead-acid and nickel-iron batteries and synchronous and induction drive motors. With its long history of electric vehicle development and the absence of a domestic source of oil, Japan seems likely to be a significant contributor to electric vehicle technology in the future.

This short history of electric cars up to 1990 has covered only the major events over the 162 years since the first vehicle moved under electric power. It stops at this stage because of the dramatic effect produced by the sales mandate issued in the autumn of 1990 by the California Air Resources Board (CARB). This required that by 1998, 2 per cent of light duty vehicle sales of each automotive company selling more than 35 000 vehicles per year in California must be zero-emission vehicles (ZEVs). The required percentage then increased to 5 per cent in 2001 and to 10 per cent

Table 2.1   Major events in the history of electric cars

1800	Volta invents primary cell and battery
1821	Faraday demonstrates the principle of the electric motor
1834	Davenport demonstrates the first electric road vehicle powered by a primary battery
1859	Planté invents the secondary cell and battery
1869	Gramme constructs the first DC electric motor of more than one horsepower
1881	Trouvé makes the first electric vehicle to be powered by a secondary battery
1885	Benz demonstrates first internal combustion engined vehicle
1887–98	Range of electric vehicles developed in Europe and the USA
1899	Jenatzy takes the world land speed record of 105.9 km/h (66 mph) and holds it for three years
1900	Equal numbers of steam, electric and gasoline vehicles compete for public acceptance
1900–12	Golden age of electric vehicles but gasoline-engined vehicles begin to dominate
1921–60	Gasoline-engined vehicles dominate completely, electric vehicles disappear
1960–90	Electric vehicles appear again in very small numbers
1990	Californian zero-emission regulations spark new activity in developing electric vehicles
1990–	Increasing numbers of electric vehicles with new battery technologies appear

in 2003, by which time the sales threshold dropped to 3 000 vehicles per year.

This requirement was sufficient to propel major manufacturers into furious activity on electric car development, and although there has been some relaxation of the original zero-emission requirement timing, in the face of the manifest impossibility of the manufacturers meeting the 1998 and 2001 requirements, a requirement for 4 per cent of all vehicles sold to be zero-emission by 2003 still remains. This would require General Motors, for example, to sell about 14 000 electric cars in California by 2003. Thus, all the major motor manufacturers continue with their developments, and it is the technology being used since 1990 and being investigated for the future that the remainder of this book is concerned with.

General References

1   WAKEFIELD, E. H.: 'History of the electric automobile', (SAE, 1994)
2   'Electric vehicles: technology, performance and potential', (International Energy Agency/OECD, 1994)

Chapter 3

Propulsion methods

The conversion of electrical power to mechanical power is fundamental in making an electric vehicle possible. The early days of electric propulsion are described in Chapter 2, and from those early days until the 1960s DC motors were always used to convert electrical energy to mechanical power at the driving wheels, since the power from the battery was immediately available in direct current form, and could be controlled by voltage switching and variation of field and armature circuit resistance and more recently by high speed on-off switching of the battery supply (the 'chopper'). Both these and more recent control methods are described in detail in Chapter 4.

Since the 1960s and the appearance of electronic power switching devices suitable for high-current applications, it has been possible to convert direct current from the battery to alternating current of variable frequency and amplitude, thus making practicable the use of both induction motors and synchronous motors. The use of these AC machines has shown considerable advantages in cost, size, weight and reliability.

This chapter describes the way these different categories of motor operate and their advantages and disadvantages when used in electric vehicles. The switched reluctance motor, the brushless DC motor and the disc motor are also described and their possible application in future electric vehicles considered.

3.1 DC motors

The principle of the DC motor is shown in Figure 3.1. In this case a simplified two-pole DC machine is shown in which a rotor (or armature) has a number of coils (represented in the diagram by a single rotor coil) wound on it perpendicular to the axis of the rotor, and this coil is free to rotate between the poles of either an electromagnet or a permanent magnet.

Electric current is supplied to the rotor coil from a DC supply through brushes on a commutator ring. This rotor will then rotate if a radial

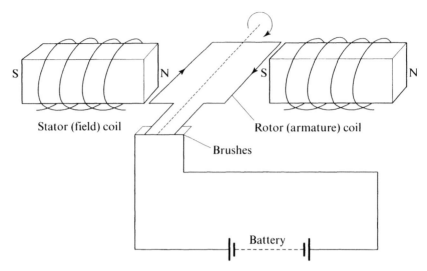

Figure 3.1   Principle of the DC motor

magnetic field exists between the North and South poles. The rotation is caused by the electromagnetic interaction between the electric field of the rotor coil and the magnetic field between the north and south poles of the stator. To maintain this interaction and the direction of rotation of the rotor, it is necessary for the direction of current flow in the rotor coil to be reversed twice for every turn of the rotor. This is achieved by connecting the rotor coils to a segmented commutator on the rotor shaft, so that the fixed brushes are connected alternately to opposite ends of the rotor coils every 180 degrees of shaft rotation.

In a more realistic embodiment of the principles described above, a two-pole DC motor would be constructed as shown in Figure 3.2. Most small or medium-size motors have two or four poles while large motors can have ten or more. Unlike an AC motor in which the number of poles is important in determining the speed, in a DC motor speed is not related to the number of poles. Where DC motors are used in electric vehicles four-pole machines are generally preferred, with poles having wound stator (field) coils rather than using permanent magnets. This arrangement is used to make it easier and cheaper to obtain the high field strength required for the few tens of kilowatts required in a lightweight electric vehicle motor. It is also used to permit electronic control of the field strength, which combined with electronic control of the armature current makes sophisticated adjustment of both motor speed and torque possible during vehicle operation.

One further advantage is also obtained by the use of a wound stator. It is easier to operate the motor as a generator in a regenerative braking mode in which energy can be fed back into the battery when the vehicle is going downhill or braking.

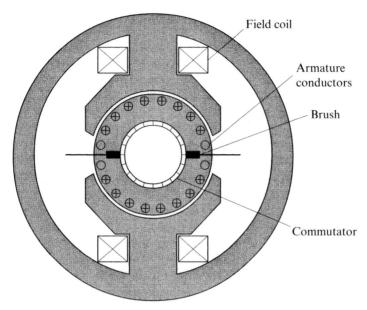

Figure 3.2   DC motor rotor and stator assembly

Prior to the availability of electronic controls it was necessary to use a DC motor with either series-wound, shunt-wound, or compound-wound field coils to obtain the motor characteristics for a particular application. The three types are shown in the diagrams of Figure 3.3 together with their torque/speed curves.

3.1.1 Series-wound motors

In the case of the series-wound motor the field coil is in series with the armature coils with the current limited by a resistor as shown in Figure 3.3(a). With this arrangement it can be seen that the series-wound motor has one major advantage when used in an electric vehicle. This is its high torque at and near zero speed. In fact, the torque would be infinite at zero speed if there was no limitation on the current available and the magnetic circuit had zero reluctance. In real life the current is limited by the series resistor R_s and the field and armature coil resistance to the maximum that the field windings, rotor windings and brushes can withstand without overheating. The torque then falls off as speed increases to a relatively small value at full speed, while the power output remains constant. The speed of the motor is most easily controlled either by adding an additional variable series resistor R_1, or adding a variable resistor R_2 in parallel with the field winding. A reduction in the resistance of R_2 then diverts current from the field winding and produces a weakening of the stator magnetic field, which in turn causes the speed to rise. An increase in the resistance of R_1 reduces the applied voltage with similar results.

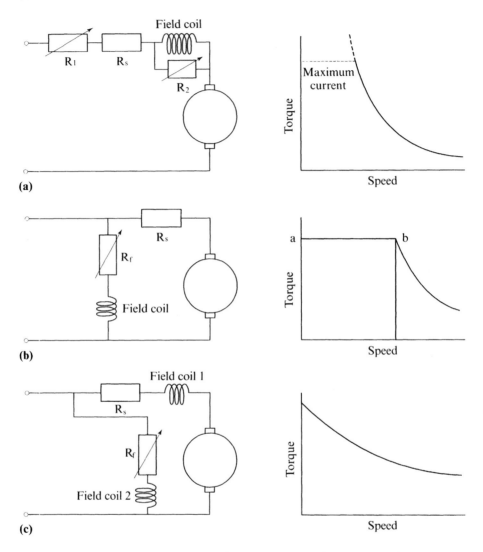

Figure 3.3   (a) Series DC motor, (b) shunt DC motor and (c) compound DC motor

The series motor characteristic is particularly suitable for an electric vehicle as it gives excellent acceleration from rest combined with a controlled slowing down on hills and a constant high speed on the flat. For this reason and the fact that simple resistive and voltage controllers could be used, series-wound motors were widely used in electric vehicles from the early days onwards.

3.1.2 Shunt-wound motors

In the case of the shunt-wound motor the field coil is in parallel with the armature coils with its current controlled by field resistor R_f, and the armature coil current limited by series resistor R_s as shown in Figure 3.3(b). With this arrangement torque remains constant as speed rises until the maximum power point 'b' on the torque-speed curve is reached, if resistor R_f is then increased in value the field current will be reduced ('field weakening') and the no-load speed will continue to increase. Under fault conditions if the field is greatly reduced or lost entirely runaway can happen, and safety devices which switch off the armature circuit if such a field loss occurs are required to avoid this possibility. Under load the speed of a shunt-wound motor remains relatively constant, dropping by only a few per cent between no-load and full-load, while the torque is directly proportional to the armature current.

Before the advent of efficient thyristor or transistor controls capable of controlling both field and armature current by turning it on and off at high speed ('chopping'), the range of speed control on a shunt-wound motor obtained by resistive field weakening was insufficient at about 3:1 to make such motors suitable for electric vehicle use. However, the ability to reverse the motor by only reversing the relatively low current field connections, instead of having to reverse the full armature current as in the series-wound motor, was a significant advantage. Since electronic controls came into use, improving the efficiency and controllability of shunt-wound motors, their use has spread although series motors still have a place in some low-cost electric vehicles.

3.1.3 Compound-wound motors

It is possible to obtain a wide range of characteristics between the extremes of the series and shunt-wound motors by combining series and shunt field coils as shown in Figure 3.3(c) in a motor having a compound-wound field. In this motor configuration the majority of the field is provided by the shunt winding, with the series field supplementing the shunt field. This is known as 'cumulative compounding'. With this arrangement the armature current in the series field increases with load causing the torque to rise but also resulting in a larger drop in speed than would be the case in a shunt motor. The characteristic can therefore to some extent be tailored to the requirement of a particular vehicle design.

3.1.4 Separately excited motors

In recent years, with the availability of power electronics with the capability of controlling high currents at relatively high voltages, it has become possible to control the armature and field currents independently in a separately

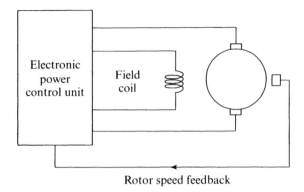

Figure 3.4   Separately excited DC motor

excited DC motor as shown in Figure 3.4. This makes it possible to obtain any required variable combination of the series, shunt and compound characteristics described in the previous paragraphs. It is essential to provide feedback of rotor speed to the electronic control system, but this is a straightforward matter. The major disadvantage that remains is the necessity to use brushes and a commutator to carry the armature current, since this limits the maximum motor speed to about 6 000 rpm and makes it not only difficult to downsize the motor, but introduces an unwelcome source of radio-frequency interference as well as wear and consequent unreliability.

It seems likely that in view of the simplicity and low cost of the control systems for DC motors and the maturity of the technology, they will remain in service for some time yet, particularly at the low-cost end of the electric vehicle market. In the longer term, however, AC motor technology has much to offer in efficiency, reliability, weight and size.

3.2 AC motors

Generally, AC motors suitable for use in electric vehicles are considered to fall into three categories: induction motors, synchronous motors and switched (or variable) reluctance motors. Induction motors and synchronous motors have been used for many years in constant-speed industrial applications but have only become possible for electric vehicle use with the advent of high-power, high-efficiency, variable-frequency inverters. Switched reluctance motors are only made possible by sophisticated electronic controls and their use and utility in electric vehicles is still being investigated. Motors using permanent magnet rotors such as the synchronous motor with rotor position feedback, paradoxically known as the 'brushless DC motor', and the double-sided axial field disc motor, are also exciting considerable interest for vehicle applications and some electric vehicles using these machines are now being built.

3.2.1 Induction motors

As in the DC motor, the induction motor develops torque by the interaction of the radial magnetic field produced by an electric current in the stator windings, and the axial current in the rotor. In the DC motor the rotor current exists because it is fed into the rotor winding from an external power source through the brushes and commutator, in the AC induction motor this current is induced by electromagnetic induction between the stator and rotor windings, hence the name. The stator winding therefore provides the magnetic field for the motor as well as the current supply for the rotor windings.

The electrical circuit of the induction motor is shown in Figure 3.5. The stator is quite different from the clearly defined individual poles of the DC motor. It consists of a steel frame with a hollow cylindrical core made of stacked laminations. These laminations have axially disposed slots distributed evenly around the circumference of the core, and it is in these slots that the field coils are wound, as shown in Figure 3.6. The rotor is also made using stacked slotted laminations, but in this case the slots may be filled either with conventional three-phase windings (a wound rotor) or with bare copper bars connected together at the ends (a squirrel cage rotor as shown in Figure 3.7). In the case of the wound rotor the coils are brought out to slip rings which turn with the rotor and make it possible to connect resistors in series with the rotor windings during starting (see Figure 3.5). Under running conditions the rotor windings are normally short-circuited, as is permanently the case in the squirrel cage rotor.

In operation the rotor is free to rotate within the stator core with a small clearance usually in the range 0.25 to 2 mm depending on the size of the motor. When a three-phase supply is applied to the stator coils a rotating magnetic field is generated within the stator core, and it is this field passing the rotor windings which induces a large current in those windings and a resultant turning force on the rotor. As the motor speeds up this current reduces because of the reduced rate at which the magnetic lines of force are

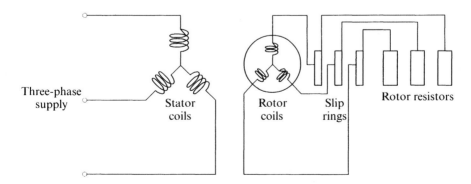

Figure 3.5   AC induction motor (wound rotor)

One turn of stator coil

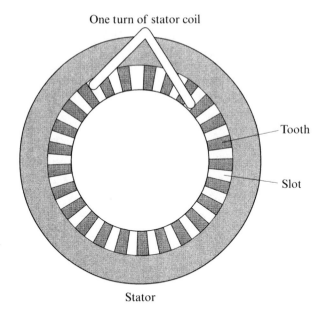

Tooth

Slot

Stator

Figure 3.6   Induction motor stator

cut until the interaction between the magnetic field from the stator and the
current induced in the rotor is balanced by the load. As long as there are
only friction and windage loads on the rotor it will rotate at almost the same
speed as the stator field, at higher loads 'slip' up to a few per cent will take
place. The rotational speed of the field developed in the stator is determined
by the number of poles, this in turn being determined by the grouping of the
stator windings in the slots (see Figure 3.8). So by increasing the number of
poles the speed of rotation of the field and therefore of the motor can be

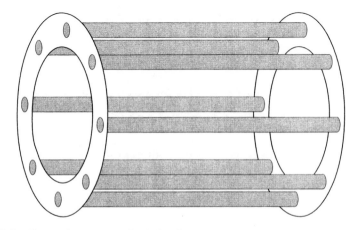

Figure 3.7   Squirrel cage rotor for induction motor

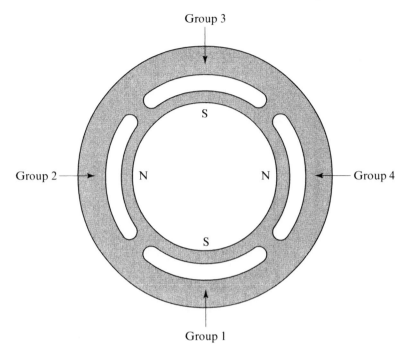

Figure 3.8   Stator field for induction and synchronous motors

correspondingly reduced. This is an important facility in industrial motors, but less so in motors for electric vehicles where it is possible to control the three-phase power frequency produced by the inverter over a wide range to control vehicle speed.

Because of the simple construction of these motors, their reasonable cost and the potential for low maintenance given by the absence of brushes and a commutator, they are attractive for use in electric vehicles. They also offer the capability of substantially higher speeds than DC motors, and since motor output is proportional to the product of torque and rotation speed, make it possible to reduce weight and size providing appropriate gearing is used. An example of this is the power/weight ratio of 0.37 kW/kg for the DC shunt motor used in the Daihatsu Hi-Jet electric car, compared to the power/weight ratio of 1.0 kW/kg for the AC induction motor used in the GM EV1, a comparable figure to the power/weight ratio for a gasoline engine.

3.2.2 Synchronous motors

The synchronous motor has many similarities to the induction motor. The stator windings are exactly the same as those in the induction motor, and when connected to a three-phase power supply, a sinusoidal rotating field is produced in the stator/rotor air gap in the same way as in an induction motor. The rotor can either have a winding which is supplied with direct current via slip rings or an arrangement of permanent magnets to produce an air gap field that matches the number of poles and sinusoidal distribution of the stator field. When the stator and rotor field patterns are aligned with North poles facing South poles, any relative displacement in either direction will result in a restoring torque trying to return the rotor to the aligned position. Under running conditions the field patterns will stay in alignment with the rotor turning synchronously with the rotating stator field. However, when load is applied to the rotor the angle between the poles of stator and rotor will increase until the restoring torque balances the load. This restoring torque will reach a maximum in a two-pole motor when the angle between the poles reaches 90 degrees. Beyond 90 degrees the rotation is unstable as the torque reduces with any further increase in the alignment angle until it becomes zero when like poles are facing each other. Wound rotor induction motors can also be made to run synchronously if they are supplied with direct current through slip rings.

Starting from a fixed frequency supply presents special problems in a synchronous motor, since for continuous torque to be produced the rotor must be rotating at the same speed as the rotating stator field. Under any other condition a pulsating torque with a zero average level will exist. To overcome this starting problem synchronous motors for fixed frequency supply use are equipped with a rotor cage similar to that used in an induction motor, in addition to either a wound rotor or a permanent magnet rotor. When the motor is switched on it operates initially as an induction motor until the rotor speed becomes synchronous with the rotating stator field, at which point it locks on and then operates as a synchronous machine. Under these conditions no current is induced in the rotor cage as there is no slip taking place. Also, since current only flows in the cage during start up, it can be rated for short-term use only. When a variable frequency power supply is available, as is the case in an electric vehicle with an electronically controlled inverter, it is possible to start the motor by increasing the frequency from a low level. This is only possible if there is feedback of the rotor position to the inverter controller so that the stator field can be positioned to match and capture the permanent magnet rotor field.

Wound rotor synchronous motors can have outputs in the range from a few kilowatts to several megawatts, whereas permanent magnet motors are limited to the range 100 W to about 100 kW. Since electric cars generally require motors in the range 5 to 50 kW, permanent magnet motors are very suitable and offer high efficiency since no power is required during running for the rotor field. As a result of this, synchronous motors have more potential for downsizing than induction motors and motors with a power/weight ratio up to 3.0 kW/kg have been built for use in electric vehicles. However, the cost of suitable high coercivity rare earth magnets and the complex electronic controller/inverters required, mean that they have not yet been widely used. Permanent magnet motors using neodymium-iron-boron (NdFeB) rare earth magnets are being used in pure and hybrid production electric cars from Toyota, Honda and Nissan.

3.2.3 The brushless DC motor

A special version of the synchronous motor is the self-synchronous permanent magnet motor with inverter, also known as the 'brushless DC motor'. In this motor the stator coils are fed from a DC source through a polyphase inverter generating rectangular current waveforms which produce a trapezoidal-shaped rotating field in the rotor/stator gap. This provides a smooth torque irrespective of speed, but requires a rotor position sensor to synchronise rotor and stator fields. This motor can be considered to be an inside-out DC machine with the rotor coils transformed to the stator position and a permanent magnet providing the field in the rotor position. No electrical connection to the rotor is required, and the characteristics looking at the DC bus feeding the inverter are those of a DC shunt motor, hence the name 'brushless DC motor'. The advantages of this motor are that it has a very high efficiency even at part load, and the rotor is of simple construction; however, the considerable cost of the high coercivity permanent magnet material required for the rotor is a major problem. In the fractional horsepower brushless DC motor the electronics for both the inverter and the rotor position sensor coding and decoding is built into the motor casing, and may even be supplied on a single chip in some cases. In the larger motors required for use in electric vehicles this is normally not possible.

An interesting variant of the brushless DC motor has been described by Hredzak *et al.* [1]. In this machine the eight-pole, three-phase stator windings are electrically conventional but are constructed in double-sided axial-field form, and the rotor is constructed as a narrow disc of permanent magnets set in a spider of non-magnetic material. The magnets, which are neodymium-iron-boron, are skewed to avoid cogging torque. The novel feature of this machine is the ability of the stator axis to move freely within limits in a perpendicular direction to the motor rotational axis. With this arrangement it is possible to attach the rotor directly to the wheel drive shaft, while the stator is attached to the vehicle chassis as shown in Figure 3.9. The vertical movement of the wheel during operation on the road then only has to move the low-weight rotor, so reducing the unsprung weight and improving the handling of the vehicle. To make this scheme practical it is necessary to measure the vertical displacement of the rotor and modulate the motor drive current to maintain constant torque irrespective of the rotor position. The problem that is likely to arise with this type of motor is the need to generate a high torque since no gearing is possible and this, in turn, is likely to require a large motor with a large amount of magnetic material.

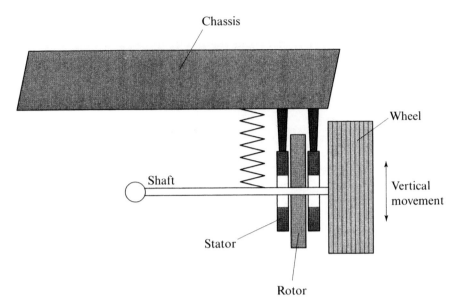

Figure 3.9   Low unsprung weight motor with rotor disc free to move with road wheel (Hredzak et al.)

3.2.4 Switched reluctance motors

The final category of electric motor with potential for use in the propulsion of electric vehicles we need to consider is that of switched or variable reluctance. It is a relatively new arrival on the electric motor scene and is probably best categorised as a special type of synchronous motor or a continuously running stepping motor. It is entirely dependent for its operation on the availability of suitable power electronics and with a control circuit similar to that shown in Chapter 4, Figure 4.5, it is possible to improve the motor efficiency significantly by recovering the stored magnetic energy when the stator coil is switched off.

The main difference between the conventional synchronous motor and the switched reluctance motor is that the stator has salient poles with the windings on each stator pole being connected so that opposite poles are connected in series.

The rotor, which has no windings or permanent magnets, is constructed of laminations designed to provide a smaller number of salient poles than those on the stator (the motor would work with a greater number of poles as well, but this is a less practical option). A typical configuration is shown in the diagram of Figure 3.10, which shows eight stator poles and six rotor poles, although other pole combinations may be used. The motor is driven synchronously by energising the opposing pairs of stator poles sequentially under the control of a rotor-shaft-mounted position detector. This causes

Figure 3.10   Switched reluctance motor – arrangement of stator and rotor

the nearest pair of rotor poles to be pulled towards the adjacent stator poles as the magnetic circuit attempts to reduce its reluctance and maximise the magnetic flux. The action is analogous to that in a solenoid. At low speeds the drive pulse in the stator windings is chopped to maintain the correct waveform, at higher speeds a single pulse is used which provides full voltage for the whole of the 'on' period of each phase. At speeds below a so-called base speed full torque is available, above base speed the current in the stator windings cannot be maintained at maximum as it is limited by the time available to apply the full drive voltage, and the torque is therefore progressively reduced as speed increases.

Because of the simplicity of the rotor design and the overall efficiency of the switched reluctance principle, this motor can offer improvements in power per unit weight, volume, reliability and flexibility in use. It does, however, suffer from uneven torque at low speed and this in turn can give problems with noise. This problem has been partially overcome in recent years by carefully shaping the current pulses to reduce the initial pull-in load on the rotor poles as each pair of stator poles is energised.

All the motors discussed are capable of operation in the 5 to 50 kW power range required in electric cars. The choice of which to use when designing a complete propulsion system is dependent on the efficiency, weight, volume and cost of the complete propulsion system of controls, motor or motors, and mechanical transmission. Comparative efficiency is illustrated by the weight comparison made in [2] for a 45 kW motor using different machine technologies:

- Wound field brush motor 130 kg
- Induction motor 80 kg
- Brushless DC motor 45 kg
- Switched reluctance 80 kg

The obvious advantage of the brushless DC machine (and the synchronous PM machine) is tempered by the cost of the magnetic components in the rotor, but it seems likely to be the motor of choice in future electric cars.

Control systems are described in detail in Chapter 4, but for this analysis it is sufficient to say that controllers for DC motors are significantly cheaper than those for the different types of AC motor already described. This cost difference is a function of the complexity of the power electronics involved but with time progress is being made on reducing this differential. It is therefore probable that AC motors will gradually take over from the still widely used DC motors, making it possible to exploit their greater efficiency and smaller size. Since a major part of that improved efficiency comes about because of the higher speed operation possible with the AC motor, it is important to include the transmission design characteristics and cost in the equation when making a decision on the propulsion system to be used.

All the electric motors described are capable of operating as generators if the control circuits are suitably designed and can therefore be used for regenerative braking in an electric vehicle. To be effective, regenerative braking must be applied over the whole range of operation of the vehicle and the mechanical brakes only used as a safety back up. When used under these conditions it is essential to avoid overheating of the motor. It is also important that the battery is capable of absorbing the returned energy at the highest required level. This may be a problem with some battery types in which case the facility to switch automatically to dynamic braking in which the energy is dissipated in resistors instead of being returned to the battery may be necessary. In energy terms it is difficult to recover much more than about 10–15 per cent of the total energy used in propelling the vehicle by means of regenerative braking, but in view of the severe limits on range in electric cars this may be considered to be well worthwhile.

The effective operation of the electric motor or motors in an electric car is critically dependent on the design of the motor control system. In the case of all of the AC motors described above the controlling power inverter and its electronic control unit must be designed to operate with the motor as an integrated system. Power inverters suitable for the different types of AC motors are described in the next chapter, together with a description of the range of semiconductor switching devices that can be used and their capabilities and limitations. The operation of the electronic control system is also discussed in this chapter with a list of the range of features that are likely to be necessary in the future.

3.3 Motor cooling

If maximum power is to be drawn from an electric motor it is necessary to provide cooling of windings on the stator and rotor and also of other vulnerable parts such as permanent magnets which may be incorporated into the motor design. Depending on the motor type, size and duty cycle this may be provided by air or a liquid coolant.

In the case of a motor for an electric car, cooling may be by air, oil or water depending on the design, with forced air cooling being the method used in most lower-rated motors. If air cooling is to be effective, ducting must be provided to get the cooling air to those components which dissipate most heat, such as stator and rotor windings. However, this necessary ducting means that the machine is larger than would otherwise be the case and there is some compromise required between improved cooling, motor size and weight. This has led to the replacement of air with water and oil. These liquids allow more effective cooling with smaller ducting and result in a motor of reduced weight and size and higher specific output.

In the case of water there is a need to avoid any contact with live parts of the motor unless deionised water is used. This restriction does not apply with oil and splash cooling as well as ducting adjacent to the electrical windings can be safely used to cool both rotor and stator, although this may as a consequence cause some viscous drag if oil enters the air gap between rotor and stator. Oil also has the advantage that the cooling function can be combined with the lubrication function particularly in a propulsion system with integral motor and gearbox. In the case of both oil and water a radiator is sometimes required to remove the heat from the cooling fluid and this may be incorporated into the vehicle heating system.

Air cooling was used by General Motors on the early EV1 (Impact) electric car motor, with a variable speed electric fan providing forced air cooling when ram air is insufficient. Later versions of the EV1 use water cooling. Ford have preferred combined oil cooling and lubrication on their ETX and NGP series of transaxles. In this system the oil is circulated by an electrically driven AC high-voltage pump with two-thirds of the oil going the motor-cooling circuit and one-third going to lubrication [3]. Pump operation is controlled by electronic logic and no external heat exchanger is required. The oil used is compatible with most plastic, elastomeric and rubber seals and does not damage the motor insulators, epoxy resins and power cable installations.

3.4 Transmission systems

In a rear-wheel-driven internal combustion-engined vehicle the drive train consists of the engine, clutch, transmission, propeller shaft, differential gears, half shafts and wheels. This complexity is necessary to convert the engine output, which can vary in speed between 800 and 5 000 rpm into the zero to 1 500 rpm speed range required at the road wheels under normal operating conditions. The drive train must also accommodate the difference in inner and outer wheel speeds during cornering, and the wide range of power output required.

In the case of the electric vehicle, although it is possible to simply replace the internal combustion engine by an electric motor as shown in Figure 3.11(a), this would not take advantage of many of the characteristics of electric drive. In particular, the ability to start from zero speed makes it possible to eliminate the need for a clutch, and the available speed range is sufficient to not require the use of transmission gears. This arrangement is shown in Figure 3.11(b). It should, however, be said that the use of gears which allow the motor to run at much higher speed for a given road speed, will add considerably to the efficiency of the complete powertrain. An attempt to realise this was made by Ford in the 1980s' development of the ETX series of transaxles, now further developed in the 1990s into the Next Generation Powertrain (NGP) Series [3]. This uses a single AC induction motor driving a two-speed automatic transaxle on a common axis, and is designed to drive two wheels differentially through a hollow motor shaft The general layout is shown in Figure 3.12(a) with a photograph of the transaxle in Chapter 2, Figure 2.11.

A further step towards simplifying the drive train and eliminating the differential can be taken by using two electric motors each connected to one of the two driven wheels. There are three different ways of doing this. The first, illustrated in Figure 3.12(b), is to mount the two motors on the vehicle chassis and connect them to their respective wheels by means of a short constant-velocity half-shaft. This makes it possible for the wheels to move freely without the weight of the motors affecting the unsprung weight of each wheel. This is important because the handling of a vehicle is critically affected by road surface inputs to the wheel and its associated parts, since these are not isolated by the suspension system as they are from the rest of the vehicle. The second dual-motor arrangement, shown in Figure 3.12(c), is to pivot each motor at one end from the chassis with the motor drive shaft directly connected to the wheels. The third option is to mount a motor in each wheel as shown in Figure 3.12(d), although this requires a motor of very high power-to-weight ratio, both because of the space available and the need to keep the unsprung weight as low as possible. This is not a new idea, as is shown in Chapter 2 where two electric cars designed in 1900 and 1902 by Ferdinand Porsche and using wheel motors are described and pictured.

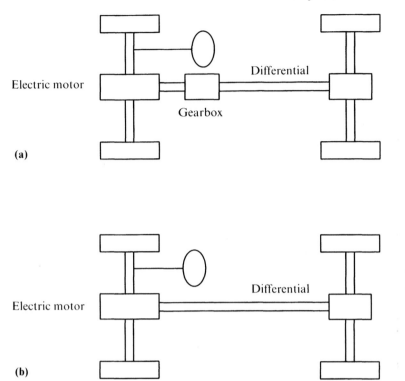

Figure 3.11 Electric vehicle drive train with the electric motor directly replacing the conventional internal combustion engine and driving the rear wheels, (a) with gearbox, and (b) without gearbox

A considerable amount of work is being done to develop motors suitable for in-wheel use, but it is a formidable task. This is because of the cost of producing the very small, high-torque, high-power motors required and the complexity introduced by the desirability of using gearing between the motor and the wheel. Cost is also a major factor in deciding if motors can be used in all four wheels, an arrangement that would give optimum independent control of traction and handling. There are also several specific problems associated with in-wheel motors; these are the increase in unsprung weight and the consequent effect on handling; the effect of heating from braking on the motor, made worse by the difficulty of providing effective cooling; and the vulnerability to damage of a motor in this exposed position.

In the current state of development of electric vehicles most practical cars use either the arrangement of Figure 3.12(a) (Ford Ecostar), or Figure 3.12(b) (GM EV1 and Nissan FEV). However, many conversions of existing conventional cars use the arrangement of Figure 3.11(b), with the electric motor replacing the internal combustion engine and its transmission directly. When developing an electric car it is essential that the motor and

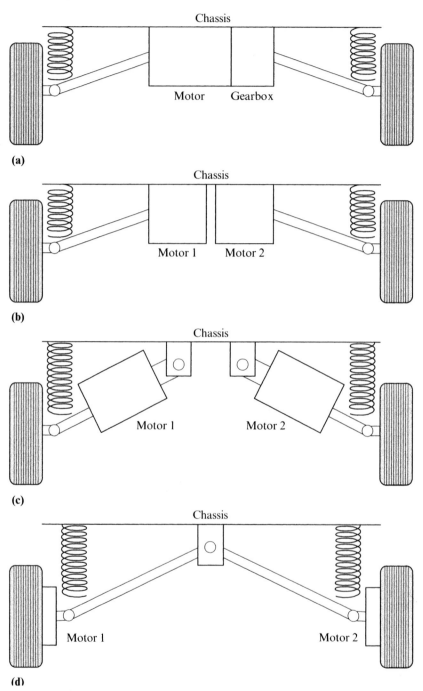

Figure 3.12   Options for electric vehicle drive trains taking advantage of the flexibility of electric motor number and position, (a) sprung single motor with integral differential and gearbox, (b) sprung dual motors, (c) partially sprung dual motors, and (d) unsprung motors in wheels

transmission design is fully integrated so that optimum cost, reliability and performance is achieved. This is particularly true when AC motors are used since it is highly desirable that their high-speed capability, and consequent low mass and volume, is fully exploited.

More complex transmission systems are sometimes required in hybrid internal combustion/electric vehicles, where it may be necessary to switch between electric only propulsion and internal combustion only propulsion, or to combine these two methods for maximum power. This is discussed in more detail in Chapter 10.

References

1   HREDZAK, B., GAIR, S., and EASTHAM, J.F.: 'Control of a novel EV drive with reduced unsprung mass', *IEE Proc B, Electric Power Applications*, 1998
2   COPUS, A.: 'Drive alternatives for electric vehicles', Battery Electric and Hybrid Vehicles Seminar, IMechE, 10–11 December 1992
3   WILLIS, R. L. and BRANDES, J.: 'Ford next generation electric vehicle powertrain', 12th Electric Vehicle Symposium (EVS), December 1994, pp. 449–58

General References

1   HUGHES, A.: 'Electric motors and drives' 2nd Edn (Newnes, 1993)
2   WILDI, T.: 'Electrical Machines, Drives, and Power Systems' (Prentice-Hall, 1981)
3   MACPHERSON, G. and LARAMORE, R. D.: 'An Introduction to Electrical Machines & Transformers' (John Wiley, 1990)

Chapter 4

Controls and power electronics

If an electric vehicle is to operate efficiently and effectively it is essential that the total vehicle system is optimised at all times to ensure that the energy available is used as effectively as possible. The amount of energy available is normally much less than that in a gasoline-powered vehicle, but the performance needs to be comparable if the electric vehicle is to operate on the road system at the same time as conventional vehicles.

In the early days of electric vehicles only the electric motor speed and torque were controlled and this was done by switching batteries in and out to give coarse voltage control and by variation of field and armature resistance of the DC motors universally used at that time. Typical basic control circuits for series, shunt and compound DC motors are shown in Chapter 3, Figure 3.3. These control techniques were adequate to make the early electric vehicles competitive, but once the internal combustion engine was fully developed in the first decade of the 20th century (see Chapter 2), the performance of vehicles using this form of propulsion was so much improved that electric vehicles ceased to be of any interest. When they appeared again in small numbers in the 1960s the early methods used for the control of DC motors were still in use. They were eventually superseded by more efficient 'chopper' techniques which permitted a much finer control of power to the motor armature than was possible by battery switching and resistance variation. Initially electromagnetic chopper technology was used, but this gave way to transistor switching as soon as transistors with adequate power ratings became available. A typical transistor power chopper circuit of the type used for the control of armature current in electric vehicle motors during the 1970s and 1980s is shown in Figure 4.1(a). The transistor T1 controls the main current supply to the armature by a pulse-width modulated voltage that is applied to the transistor base and is derived from the position of the drivers accelerator pedal. Diode D2 provides a path for the armature current when T1 switches off. Figure 4.1(b) shows a further development of the circuit that by the addition of a second transistor T2 and diode D2 makes it possible to achieve regenerative braking by switching on

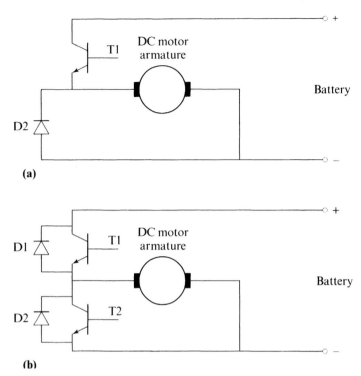

Figure 4.1   (a) Transistor power chopper for armature current control, and (b) addition of second transistor to permit regenerative braking

T2 and feeding energy back to the power source when the brake pedal is pressed.

Many of these simple control systems are still in use, but in recent years it has been recognised that if electric vehicles were to fully exploit their zero-emission advantage they would have to compete more effectively on performance with conventional vehicles. To achieve this it was clear that every aspect of the vehicle would have to be controlled by a sophisticated electronic energy management system using complex software [1] if the limited energy available was to be used in the most efficient way possible.

4.1 Electronic energy management

A block diagram of a comprehensive energy management system for an electric vehicle using one of the electrochemical battery types described in Chapter 5 is shown in Figure 4.2 with the likely inputs and outputs indicated. Other energy storage methods such as those discussed in Chapter 6 would require some changes to the input and output parameters shown. The microprocessor control system makes use of a range of inputs from sensors,

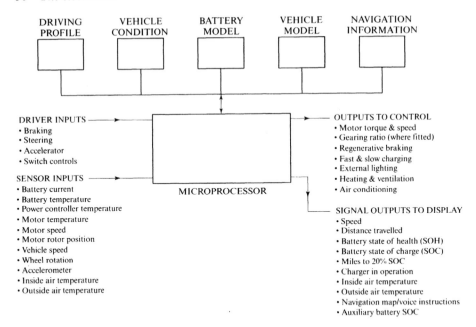

Figure 4.2   Energy management system

measuring battery, motor, vehicle and ambient conditions, and combines this information with driver-demand inputs from braking, steering, accelerator and the various switch controls available. Then, using electronic models of the vehicle and the battery held in memory and optimising for the best energy usage, outputs are generated by the microprocessor to continuously control motor torque and speed, gearing ratio (where gearing between motor and drive wheels is used), regenerative braking, external lighting, heating, ventilating and air conditioning. When the vehicle is stopped and plugged into a charging station, the microprocessor will monitor the battery, generate the charging algorithm and control the charger. In the more sophisticated systems navigational information can also be held in memory and processed to provide navigation instructions to the driver. A further useful feature is the ability to generate information on the vehicle and battery condition and the way it is being driven, and to hold this information in reprogrammable memory. This enables the driver to obtain information on the distance remaining before the battery will require recharging if he continues to drive in the same way, and on any functional problems with the vehicle. The system also provides information for the driver instruments showing speed, distance travelled, state of charge, miles to battery 'empty' (normally considered to be at 20 per cent state of charge (SOC), charger in operation and inside and outside air temperatures.

The configuration and complexity of both the electronic controller and the power electronics in any control system are affected by a number of factors, not least of which is the number of motors to be used. As can be seen from Chapter 3, Figures 3.11 and 3.12, one, two, or in the case of in-wheel motors, four are possible. More than one motor effectively excludes the use of gear changing as a method of optimising efficiency, as the complexity is too great. The use of two drive motors as shown in Figure 3.12(b) and (c), requires the use of separate power driver circuits and separate fixed ratio planetary gears for each motor. This ensures that it is possible to adjust the torque between the two motors when the vehicle is cornering so that it is reduced on the inner wheel and increased on the outer. There is also a potential problem if power is lost to either motor as the vehicle could veer to one side and the control system must be programmed to take care of this, since it is a significant safety issue.

Another factor determining the design of the control system is the type of motor or motors to be used. The DC motor is the easiest to control since rotor-position sensing is not needed, but if a series motor is used it is necessary to make provision for reversal of the series field when reverse rotation is required. The induction motor requires closed-loop vector control of flux by means of rotor-position sensing and a complex control algorithm. In the case of the synchronous motor closed-loop control is needed with an encoder feeding rotor-position information back to the inverter controller. If the brushless DC motor is used a rotor-position sensor is required to synchronise rotor and stator fields. Finally, the switched reluctance motor can operate with relatively simple algorithms, but the complexity of the algorithm is greatly increased if torque ripple is to be reduced to improve motor noise, and efficiency is to be maximised. Other factors affecting control-system complexity include the use of a gearbox which requires electronic control if energy transfer between the motor and the road wheels is to be optimised, and control of the power electronics switching when reverse rotation of the drive motor or power regeneration during braking is required.

The use of a high battery-system voltage can have a significant effect on the power electronics since current is reduced and losses due to voltage drops in the power elements are reduced. However, safety considerations limit the voltages used. The range of voltages used in most electric cars lie between 200 V and 350 V, although there have been proposals to use over 500 V for special vehicles.

The comprehensive energy management system will need to control all the auxiliary systems in the vehicle including lighting, de-misting, de-icing and seat heating. These will operate from a much lower voltage than that of the main battery both for safety reasons and to permit standard components to be used. At present, a 12 V auxiliary battery would be used charged through a DC/DC converter from the main battery. Future auxiliary systems may require a 42 V power supply of the type now under consideration for the conventional vehicles of the next few years. This low voltage operation will also be used for all the small motors and solenoids used around the vehicle for door locking, window opening, seat adjustment and other convenience functions. The air-conditioning compressor will, however, operate at full main battery voltage to avoid the conversion losses that would occur if such a high power system were operated at low voltage.

4.2 Power electronics

Where DC motors are still used for propulsion they are normally of the separately excited type shown in Chapter 3, Figure 3.4, it being possible to obtain any required variable combination of series, shunt and compound characteristics by controlling the armature and field currents independently from a suitable electronic control unit. It seems likely that DC motors will continue to be used at the low-cost end of the electric vehicle market because of their simplicity and low cost, but if efficiency is to be maximised then the trend towards AC drive systems will continue and eventually dominate the market.

Currently, AC drive systems usually use three-phase induction motors. These are driven by a pulse-width-modulated inverter having the circuit configuration shown in Figure 4.3.

In this circuit six power modules are used in a three-phase bridge connection to form the inverter. Each power module may contain a number of

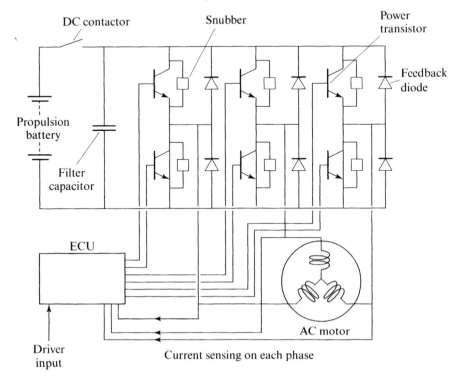

Figure 4.3   Induction motor drive circuit

power transistors connected in parallel to give sufficient current capability to meet the requirements of the system (in the diagram each group of transistors in a power module is represented by a single device). Across the transistor is also connected a parallel fast feedback recovery diode to provide a return path for the motor current when the power transistor switches off.

The current to the motor is controlled by pulse-width modulation (PWM) of each of the six power module transistors with control signals of suitable width, frequency and phase to optimise the alternating current sent to each phase of the motor. The pulse-width modulation is primarily controlled from the induction motor torque, this being derived by measuring the stator current and inferring air gap flux from the motor terminal voltage and then controlling the system to produce maximum load torque for minimum peak current.

A snubber to control the shape of the switching waveform may be connected across each power transistor module and consists of a resistor-capacitor circuit in parallel with the power module, together with a small parasitic inductance in series with the power module as shown in Figure 4.4. If parallel power transistors are used, then they are matched so that each one can be constrained by the snubber to work within its safe operating area (SOA).

A filter capacitor bank is connected across the input from the propulsion battery, both to filter the DC input voltage and to provide a low impedance path for the high frequency currents generated during pulse-width modulation switching. Ideally this capacitor should be large enough to reduce the resonant frequency of the power source inductance and the supply capacitance to well below the switching frequency of the controller, which is likely to be in the range between 1 to 2 kHz. It is possible to accomplish pulse-width modulation, with a small efficiency loss, by switching the inverter transistors at high frequency and varying the duty cycle at the desired motor frequency. This is of help in increasing the system resonant frequency and also permits sine-triangle, or sine and third-harmonic waveform shaping, making it possible to obtain a higher specific output from the motor.

One phase of a typical inverter power transistor module could therefore consist of four parallel matched power switching transistors of 500 V to 600 V rating, carrying a maximum current between 150 A and 200 A with the variation in current sharing between them controlled to within 30 A by the use of a small series resistor in the base circuit of each transistor. A snubber would also be connected across the transistors and to isolate any transistor that develops a fault during operation a blocking diode would also be included.

This general design of a three-phase bridge inverter can be used for both synchronous motor and brushless DC motor drives as well as for induction motors.

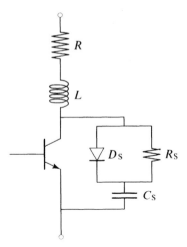

Figure 4.4   Snubber circuit

Another candidate for AC electric vehicle drives is the switched reluctance (SR) motor. The motor technology is described in detail in Chapter 3. It uses a control system that switches the stator pole pairs in sequence to produce a rotating field that attracts the rotor poles towards each stator pole pair in turn, giving a synchronous stepped operation. This motor can be considered as a stepper motor, but should be considered in this context as an AC machine.

The control system for a SR machine requires microprocessor technology to ensure that stator currents are switched according to a stored map of the optimum switching points. The inverter circuit for a SR motor is shown in Figure 4.5 and contains flywheel loops as shown to permit the current to continue to flow in the load when the power control transistor is switched off [2]. This gives a relatively smooth current load on the battery.

The timing and wave shape of the switching currents supplied by the inverter need to be carefully controlled if motor noise caused by torque ripple is to be minimised. Continuing development is taking place to resolve this problem and to fully develop the SR drive's unique advantages of high efficiency and high torque.

There are a number of different types of power transistor suitable for use as power switches in the controllers that have been described and these will be described in the next section.

4.3 Power switching devices

Until the end of the 1980s the two-stage Darlington transistor was first choice for AC drive systems. However, since then other semiconductor switching devices such as MOSFETs and IGBTs have taken over. These are

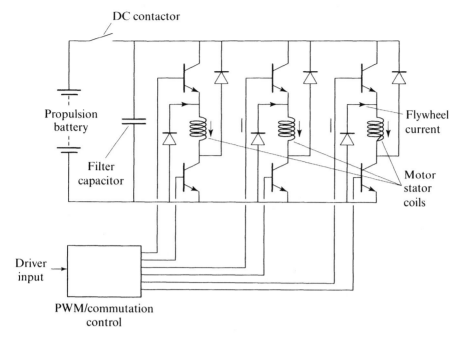

Figure 4.5   Switched reluctance motor drive circuit

capable, either individually or in parallel-connected assemblies, of control-ling drives up to the 50 kW or so required for a small passenger vehicle. Devices need to have ratings of at least 600 V and in the appropriate power module assemblies be capable of handling battery supplies in the 200–300 V range together with the inevitable switching transients.

Drive motor currents up to 500 A can be required when operating under high-load, low-speed conditions when the full peak rating of the drive motor may be used. Power module designs must allow for short periods of operation under these conditions.

The various power switching devices currently available or under devel-opment will now be described in more detail.

4.3.1   The bipolar Darlington

Bipolar junction transistors (BJTs) have the potential for a very low collec-tor-emitter saturation voltage drop under the high-current, high-voltage conditions required for electric vehicle drive applications. They do, however, require very high base drive currents if this very desirable characteristic is to be realised. These very high base currents can be provided most conve-niently by the use of the two-stage Darlington configuration shown in Figure 4.6, and this arrangement is used in low-cost power-switching systems.

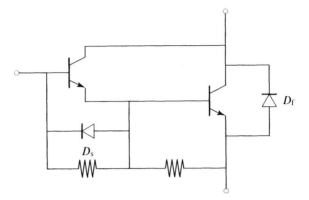

Figure 4.6   Two-stage bipolar Darlington circuit

Two-stage Darlington power transistors are readily available and can be paralleled by the use of a balancing resistor. It is essential to use this method of balancing as the on-resistance of bipolar transistors decreases with increasing temperature, and if the current is not controlled, current localising can occur internally in the device, leading to destructive failure. At low temperatures the base current has to be increased if the saturation voltage drop is to be kept to the level existing at ambient temperature, although self-heating of the device under these conditions will improve the situation.

Three-stage bipolar Darlington transistors have been developed and require lower base drive currents than their two-stage counterparts, but the collector-emitter saturation voltage is increased and the advantages over the two-stage device have not been sufficient to compensate for the higher cost and the possible need for more paralleled devices.

4.3.2 The thyristor

When first developed, the thyristor (sometimes known as the silicon controlled rectifier (SCR)) was seen as a major advance in making it possible to control large amounts of power, since it was capable of handling much higher currents than the relatively low-power bipolar devices then available. It also had a much lower saturation voltage drop than the then available bipolar devices. This, combined with its ability to be turned on by a short current pulse and then not require any further control current input to maintain the current flow through the device, was seen as giving it a significant advantage. However, the thyristor could only be turned off by stopping current flow through the device for the few microseconds necessary for it to revert to the off state ready for the next turn-on current pulse. This requirement limits the use of the thyristor to relatively low-frequency switching systems (below 1 kHz). Turn-off is normally achieved by using an appropriately charged commutation capacitor to divert the main current flow from the device for a short time. However, this arrangement is susceptible to unplanned turn-on by electrical noise, a condition that could be damaging to the system.

Because of the thyristor's high current-carrying capacity and 1 000 V+ operating voltage it was widely used in electric vehicle controllers in the 1970s, but has since been successively replaced by the bipolar Darlington and then by MOSFETs or IGBTs. However, there have been further developments in thyristors intended to make them more attractive for power control and which enable them to be turned off through a control gate on the device. Thyristors with this facility are known as gate turn-off thyristors.

4.3.3 The gate turn-off thyristor (GTO)

The equivalent circuit of the GTO thyristor is shown in Figure 4.7 and is configured as two regeneratively coupled transistors. When a positive voltage is applied to the base of the NPN transistor to drive a turn-on gate current into the base, electrons are injected from the cathode into the base region. These electrons diffuse through the base of the upper NPN transistor to the collector and act as a base current for the PNP transistor. If the sum of the current gains of the two transistors exceeds unity, the device latches into its forward conducting state. To achieve turn-off, a negative gate current must be applied. This gate current must be sufficient to lower the NPN transistor gain until the sum of the two transistor gains is less than unity; this will then extinguish the forward-current flow.

Unfortunately, the above process results in an order of magnitude increase in the turn-on current and a significant increase in forward-voltage drop compared to the very low level of a conventional thyristor. It may also be necessary to maintain gate drive during conduction if an acceptable forward voltage drop is to be maintained.

The GTO thyristor retains the high-voltage capability of the original thyristor and because of the shorter turn-off time is capable of operating at

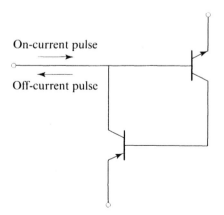

Figure 4.7   Equivalent circuit of gate-turn off (GTO) thyristor

significantly higher switching frequencies. It is also suitable for use instead of the thyristor where the features of high-frequency operation and easier turn-off are required. It is, however, a significantly more expensive device than the thyristor.

4.3.4 The MOS-controlled thyristor (MCT)

In this integrated device, a metal oxide silicon field effect transistor (MOSFET) is incorporated across the base-emitter junction of either a NPN or PNP transistor. This combination produces the bistable characteristics of the thyristor. Shorting out the relevant base-emitter junction by turning on the MOSFET interrupts the positive-feedback path, thus turning off the MCT. Apart from its turn-off capability, the device is reported to be suitable for high-voltage operation and to have fast switching speed, high power handling ability, superior dynamic characteristics and high reliability [1]. With these characteristics it should prove a very useful switching device for electric vehicle power electronic control systems.

4.3.5 The MOSFET

The power metal oxide silicon field effect transistor (MOSFET) was first developed in the 1970s to meet the demand for power transistors which could be controlled with a much lower current than the bipolar devices then available.

A power MOSFET blocks current flow in the absence of gate bias. If a positive voltage is applied to the gate, electrons are attracted to the surface of the base region so that when the bias exceeds a certain level, a conductive channel is formed which can sustain a source to drain current. The 'on' resistance increases as the 2.5 power of the blocking voltage and this high rate of increase limits the use of these devices to less than 1 000 V. This 'on' resistance also increases rapidly with temperature so that it is necessary to limit operation to less than 400 V if a reasonable 'on' resistance at high current is to be obtained.

One advantage of the increase in 'on' resistance with temperature is the ease with which MOSFETs can be operated in parallel, since currents tend towards balance between paralleled positive temperature coefficient devices. MOSFETs also have very fast switching speeds of more than 10 MHz.

MOSFET costs are higher than those for bipolar junction transistors, especially when their relatively low voltage capability and the need to operate a number in parallel to obtain the current output required for electric vehicle operation is taken into account. The general consensus is that these devices cannot effectively meet the combination of low 'on' resistance and high-voltage capability required for electric vehicle propulsion motor drives. These parameters are more cheaply and easily met by the use of an insulated gate bipolar transistor which will be described in the next section.

4.3.6 *The insulated gate bipolar transistor (IGBT)*

The IGBT is a power-switching device which combines power MOSFET and bipolar device technology to try to obtain MOSFET control characteristics with bipolar output characteristics. The device is designed to combine low forward-voltage drop with fast switching speed at high-current density to produce the ideal power switch. The equivalent circuit is shown in Figure 4.8.

When a positive voltage is applied to the collector with the gate biased at emitter potential, the device operates in its forward-blocking mode; but if a large enough positive gate-to-emitter voltage is applied, the device switches to its forward-conducting state and operates at high current density in the same way as a forward-biased diode. The device has a gate-to-emitter threshold voltage and a capacitive input impedance, so that in order to turn the device on the input capacitance must be charged to the threshold voltage before the collector current can begin to flow. To turn the IGBT off, it is necessary to provide a path for the input capacitance to discharge. If switching rates are to be high, then high input and discharge currents may be necessary, although only small currents are needed to maintain the 'on' condition.

Operation at up to 600 V with a forward-voltage drop of about one-seventh of that for a power MOSFET and a current capability which exceeds that of a bipolar transistor is possible. IGBTs can be connected in parallel although gate-to-emitter threshold voltages and collector currents

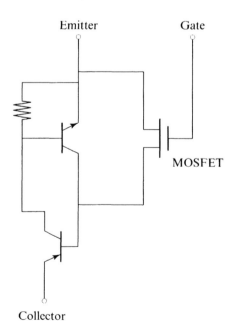

Figure 4.8   Equivalent circuit of insulated gate bipolar transistor (IGBT)

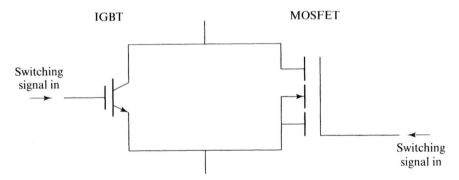

Figure 4.9   Parallel operation of IGBT and MOSFET

must be balanced. For higher currents the temperature coefficient is positive so that the IGBT behaves in a similar way to MOSFETs in parallel under these conditions. High switching rates up to 66 kHz with 50 A peak current at 500 V DC have been reported [3] without the use of a snubber. For higher currents and higher switching frequencies, a snubber is required.

A further development of IGBT technology is to parallel the IGBT with a MOSFET, as shown in Figure 4.9. This arrangement overcomes the relatively large loss during turn-off in an IGBT due to the tail current of the bipolar output stage. By turning the IGBT part of the pair off first and allowing the load current to flow through the MOSFET, the IGBT switching losses are avoided. The MOSFET can then be switched off. It does, however, require sophisticated switching control to ensure that the division of load current between the IGBT and the MOSFET is optimised for best overall efficiency.

The characteristics of all the devices described are summarised in Table 4.1. Both IGBT- and MCT-based inverters are now being used in electric vehicles developed by the major manufacturers [1] with substantial gains not only in operating characteristics but also in inverter weight and volume.

Other components also play critical roles in making the power electronics system work effectively and these will be described in the following sections.

4.4 Semiconductor cooling

In an electric road vehicle application the power dissipated by the semiconductor components is normally considerable and makes a substantial heatsink essential. The voltage drops across the semiconductor devices when conducting will therefore be an important factor in the selection of suitable devices. It should be noted that in an inverter for an AC drive it is usually the case that current from the supply must pass through two devices in series rather than through the single device required in a DC controller. Thus, with the same voltage drop across each switch when it is

Table 4.1   Characteristics of power switching devices

Transistor type	Maximum blocking voltage (V)	Maximum steady current (A)	Maximum operating frequency (kHz)	Gate drive power	Conduction drop	Ease of paralleling	Cost
Two-stage Darlington	1 400	800	20	High	Low	Medium	Low
Thyristor (SCR)	3 000	2 000	1	N/A	Low	Not required	Low
GTO Thyristor	3 000	2 000	10	Medium	Medium	Not required	High
MCT	1 000	150	100	Low	Low	Medium	High
MOSFET	1 000	100	10 000	Low	High	Easy	High
IGBT	1 200	400	100	Low	Low/Med	Easy	High

conducting, the losses in an AC inverter will be significantly higher than those in a DC controller.

Any heatsink must, of course, be chosen with the 'worst case' power circuit losses at the maximum specified ambient temperature. There are several ways in which the necessary cooling can be achieved. Natural air cooling, forced air cooling, or liquid cooling could be used, the size of the heatsink being decided by the efficiency of each method. With the second and third methods any failure of the fan or pump in the cooling system will result in a rapid increase in heatsink temperature. With liquid cooling, if the cooling system is to be self-contained some form of liquid-to-air heat exchanger (radiator) will usually be required. Also, if there is any voltage difference between individual heatsinks or between them and other parts of the cooling system, the cooling system liquid will have to be non-conducting. In practice, it is not necessary to provide heatsink capacity for continuous operation under maximum-power conditions, since the average cooling required over a vehicle driving cycle is usually relatively low and by providing adequate thermal inertia in the cooling system, the rise in temperature of the heatsinks can be kept to an acceptable level.

An alternative to conventional air- or liquid-cooling systems, which require large heavy fins in the case of air cooling, or a heat exchanger in the case of water cooling, is a cooling system making use of the large heat transfer occurring when water is boiled and then condensed. The system operates by allowing the water to boil on the outer surface of the IGBT module heatsink so that the resulting steam ascends and condenses on the inner wall surfaces of lower temperature radiation tubes. This heat is then transmitted from the radiation tubes to radiation fins that are cooled by fans that remove the heat to atmosphere. The condensed water is returned by gravity to the heatsink surface to be recycled through the system. It is claimed that a cooling system using this technology can reduce the cooling components in a 100 kVA inverter by 32 per cent in volume and 27 per cent in weight compared to a conventional water-cooling system [4].

Since the power devices are electrically connected to the battery they need to be isolated from the vehicle chassis. This leaves two possible ways of mounting the power devices on their heatsinks: they can either be electrically connected to the heatsinks and isolated from the adjacent metalwork, or isolated from the heatsinks which are then connected to the metalwork and can safely be provided with cooling fins exposed to the airflow around the vehicle. The first option permits the use of non-isolated devices, but to keep down internal temperatures, forced cooling may be required, particularly if the controller is totally enclosed. The second option provides a good solution if the devices can be easily isolated. Power modules with electrical connections isolated from the mounting base are particularly suitable. Non-isolated devices require insulation to be included between device and heatsink. Preferably such devices should be directly mounted on metal blocks insulated from the heatsink by thin plastic film used in a double layer to reduce the risk of pinholes causing shorting. The metal blocks spread the heat from the devices, increasing the area of film through which heat is conducted to the heatsink and reducing the temperature drop across the film. Care must be taken with the designs using this method of isolation if problems in unfavourable environmental conditions are to be avoided.

4.5 Capacitors

Capacitors are critical components in AC power circuits, particularly those in the filter capacitor bank and snubber used in the power electronics circuits described in Section 4.2. Selection of an appropriate type is therefore very important. Electrolytic capacitors can provide high capacitance per unit volume, but both cost and physical volume are too high for devices with acceptable ripple-current rating and series resistance. Electrolytic capacitors also degrade as they age and are adversely affected by low temperatures.

Capacitors using plastic film dielectric with metallised electrodes are an alternative that offers high ripple-current rating, low effective series resistance and the capability of self-healing, so giving high reliability. However, the capacitance available is limited if cost and physical volume are to remain acceptable. For this reason circuits must be designed and operating conditions arranged so that satisfactory performance can be achieved with relatively low values of capacitance.

4.6 Current measurement

One of the most critical measurements that has to be made if the control system of the electric vehicle is to operate effectively to control motor speed, reversing and regenerative braking, is that of power-circuit current.

The simplest method is to place a shunt resistor in series with the current to be measured. It responds to both AC and DC without the need for an additional power supply. To avoid significant power dissipation it has to be of a very low resistance, and consequently will produce a very small measurement voltage under low-current conditions. If the resistance is increased to obtain a higher voltage output, power loss will be increased and an error in the measurement may occur due to the shunt resistor self-heating. Alternatively, an amplifier may be needed which will have to be electrically isolated and designed to be resistant to the very high level of electrical background noise to be expected in high power switching circuits. The inductance of the shunt can be sufficient to distort the waveshape across it under rapid-switching operation and large spikes can be generated. Very low inductance coaxial shunts are an alternative but they are bulky and expensive.

For AC power electronic systems it is possible to use a transformer. This provides effective isolation between the conductor in which the current is being measured and the measuring system. If the transformer is designed to have minimum stray capacitance and leakage inductance then it will have a short response time. With the use of a suitable primary-to-secondary turns ratio it is possible to obtain a high output level with a good signal-to-noise performance; it is, however, necessary to take particular care with the disposition of the windings to minimise interwinding capacitance and any consequent common-mode interference.

Another alternative is the Hall-effect sensor, which can be used to measure the magnetic field around a conductor and is suitable for measuring both AC and DC. This sensor does not introduce any losses into the circuit being measured and is effectively isolated from it; an amplifier is required, however, if a high output is needed. There may be some output drift with temperature and time and compensation for this will need to be included. A stabilised power supply will also be required and this, together with the amplifier and compensation circuits, makes the complete Hall-effect sensor system a relatively expensive solution.

One other effective but relatively expensive method for measuring current combines a transformer to provide AC measurement together with a Hall-effect device to maintain measurement down to DC. Known as a current-measuring module, it is available commercially and provides good isolation, good speed of response and minimal losses in the circuit being measured. Minor output changes can occur with temperature and time but these effects are small. A low-impedance output giving good noise immunity is usually provided.

Of the current-measuring devices described, the shunt resistor still provides the cheapest solution for many applications. Only when isolation or a large signal level is necessary are the other devices required, with the transformer being most suitable for AC only applications. The current module is ideal for more sophisticated systems where cost is not a major constraint.

In combination, the devices and systems described above can provide sophisticated and efficient power electronics for electric vehicles, and with high-quality software engineering, optimised operation of energy management systems of the type shown in Figure 4.2.

The principal task of designers in the next 20 years is to reduce the cost of these systems by integration and improved manufacturing and assembly methods so that low-cost controllers are available to contribute towards the essential reduction in electric vehicle cost required for their success in the market place of the future.

References

1   CHAN, C. and CHAU, K.: 'An overview of power electronics in electric vehicles', *IEEE Transactions on Industrial Electronics*, February 1997, **44**, (1)
2   APPLEYARD, M.: 'Electric vehicle drive systems', 8th International Automotive Electronics Conference (IEE London, October 1991)
3   JOHANSON, J., JENSET, F., and ROGNE, T: 'Characterization of high power IGBTs with sinewave current', *IEEE Transactions on Industrial Applications*, September/October 1994, **30**, (5)
4   ISHII, J., FURATA, S. KAWAGUCHI, K., and SUZUKI, M: 'Inverter with a new cooling apparatus by boiling and condensation', 13th Electric Vehicle Symposium (EVS), Osaka, Japan, October 1996, pp. 567–73

Chapter 5

Energy sources 1 – Storage batteries

Over the 160-year history of the electric vehicle, the capability of the battery to store energy has always been the factor limiting the success of this form of propulsion. In the period 1900–1910, when electric and gasoline propulsion were competing for the burgeoning domestic and business transport market, it was the restricted range and slow refuelling of the electric car which finally resulted in the domination of the market by the internal combustion engine.

The restricted range of the electric car is caused by the limited amount of energy which can be stored in a practical battery, and this is better appreciated when it is understood that a conventional lead-acid traction battery has an energy density of about 35 Wh/kg compared to a useful energy density of more than 2 000 Wh/kg for gasoline.

It should also be understood that there are many materials and chemicals which can be combined in couples to make viable primary and secondary battery cells. Because of the large number of different sources of energy that are already in use or are projected for use in electric vehicles in the future, this subject will be covered in two chapters. In this first chapter we shall only consider chemically based electric secondary storage batteries, while in the second chapter we will describe other means of storing and generating energy which can be used to power electric vehicles directly or can be used to augment battery power.

Table 5.1 lists the storage battery types which are currently either already being used in electric vehicles, or are under consideration for use in the next decade. The maximum energy density column shows the energy storage capability of each battery type in Wh/kg of battery weight at a three-hour discharge rate. This figure indicates the storage capacity of the battery and how many Watt-hours of energy can be extracted for each kilogram of battery weight when discharged over a three-hour period. In the case of an electric vehicle it therefore determines how far that vehicle can travel before the battery is fully discharged, given a knowledge of the number of Watt-

Table 5.1   Electric vehicle batteries (properties)

Battery type	Maximum energy density (Wh/kg)	Maximum power density (W/kg)	Fastest 80% recharge time (min)	Operating temp.	80% discharge cycles before replacement	Estimated large-scale production cost ($ per kWh)
Lead-acid	35	150	No data	Ambient	1 000	60
Advanced lead-acid	45	250	No data	Ambient	1 500	200
Valve regulated lead-acid	50	150+	15	Ambient	700+	150
Metal foil lead-acid	30	900	15	Ambient	500+	No data
Nickel-iron	50	100	No data	Ambient	2 000	150–200
Nickel-zinc	70	150	No data	Ambient	300	150–200
Nickel-cadmium	50	200	15	Ambient	2 000	300
Nickel-metal hydride	70	200	35	Ambient	2 000+	250
Sodium-sulphur	110	150	No data	350 °C	1 000	150
Sodium-nickel chloride	100	150	No data	300 °C	700+	250
Lithium-iron sulphide	150	300	No data	450 °C	1 000	200
Lithium-solid polymer	200	350	No data	80–120 °C	1 000	150
Lithium-ion	120–150	120–150	<60	Ambient	1 000+	150
Aluminium-air	220	30	No data	Ambient	No data	No data
Zinc-air	200	80–140	No data	Ambient	200	100

Note: *Useful* energy density of gasoline > 2000 Wh/kg

hours per kilometer required to move the vehicle at an acceptable speed (an Escort-sized vehicle requires about 200 Wh/km of travel on average).

The maximum power density column indicates how rapidly power can be drawn from the battery and therefore the maximum current that can be supplied to the drive motor to accelerate the vehicle. It is, however, important to understand that high rates of battery discharge reduce available energy density significantly in most battery types and therefore reduce vehicle range.

The next column shows the fastest full recharge time. This is the shortest time in which the battery can be recharged from an 80 per cent discharged condition to a fully charged state. It is very important in determining the viability of a battery in an electric vehicle where the usage may include long trips with recharging every time the battery charge falls to the 80 per cent depth of discharge (DOD) condition.

Operating temperature is an important factor in determining the ease with which batteries can be used, with those operating at ambient temperature only requiring conventional packaging and protection. Those that operate at elevated temperatures – in particular sodium-sulphur, sodium-nickel chloride, and lithium-iron sulphide – require special packaging and protection to retain the maximum amount of heat when the vehicle is parked. It is also necessary to provide battery-heating facilities for maintaining heat levels during periods of extended non-use, since permanent damage can be caused if the electrolyte freezes, a situation which can arise in the case of sodium-sulphur if the temperature drops below 200 °C.

Battery life is given by the next column, which lists the number of 80 per cent deep discharge cycles before battery replacement is required. A suitable target for a practical battery would be 1 000 cycles. This would give a useful life of about three to four years, although it should be understood that for most battery types each deep discharge cycle reduces both the available energy density and maximum power density of the battery by a small amount. For this reason, performance towards the end of the useful life of the battery is significantly reduced. Operation at low temperatures (below 0 °C) can also reduce battery performance if it is designed for operation at normal ambient temperature.

Finally, the table lists the author's best estimate of large-scale production costs for each battery type, the costs of experimental batteries being many times higher than the figures given.

Table 5.2 gives the cell voltages of the batteries listed in Table 5.1 together with the anode, cathode and electrolyte materials normally used. Cell voltage is important not only because generally the higher the voltage the better the ratio between the active components in the cell and the passive containing material, but also because voltage determines the number of cells required to assemble a battery of any given voltage and therefore its complexity and potential reliability. The lithium-ion cell shows up particularly well against this criterion with a voltage about twice that of most other cells.

A major influence on the increased pace of development of electric storage batteries over the last few years has been the activities of two consortia: the US Advanced Battery Consortium (USABC) and the Advanced Lead-Acid Battery Consortium (ALABC).

The US Advanced Battery Consortium (USABC) was formed in 1991

Table 5.2   Electric vehicle batteries (voltages and materials)

Battery type	Open-circuit cell voltage	Anode material	Cathode material	Electrolyte composition
Lead-acid	2.1	PbO_2	Pb	H_2SO_4
Nickel-iron	1.2	Ni	Fe	KOH
Nickel-zinc	1.7	Ni	ZnO_2	KOH
Nickel-cadmium	1.2	Ni	Cd	KOH
Nickel-metal hydride	1.23	Ni	Metal hydride	KOH
Sodium-sulphur	2.1	S	Na	βAl_2O_3
Sodium-nickel chloride	2.1–2.2	NiCl	Na	βAl_2O_3
Lithium-iron sulphide	1.75–2.1	FeS_2	LiAl or LiSi	LiCl/KCl
Lithium-solid polymer	2.0–2.5	Li	V_6O_{13} + acetylene black	$(PEO^*)_{12}$ $LiClO_4$
Lithium-ion	3.6	Carbon intercalation	$LiCoO_2$	Organic
Aluminium-air	1.5	Al	O_2	KOH
Zinc-air	1.65	Zn	O_2	KOH

* Polyethylene oxide

and involves the major USA auto manufacturers, the US Department of Energy (DOE), the Electric Power Research Institute (EPRI), and the battery manufacturers. It has established the cost-performance criteria for advanced batteries shown in Table 5.3, and awarded a number of contracts for development of advanced batteries to meet these criteria. Current development is concentrated on nickel metal hydride, lithium ion and lithium polymer batteries. Work on sodium-sulphur has now been discontinued.

The Advanced Lead-Acid Battery Consortium (ALABC) was formed in 1992 by pooling the R&D resources of the world's lead producers, battery manufacturers, component suppliers and related industrial sectors, on the basis that further substantial development of the lead-acid battery was possible.

We will now consider in detail the battery types listed in Table 5.1. The batteries will be considered in groups: first, four types of lead-acid battery; then four different nickel-based batteries; then, four-high temperature batteries; and, finally, the longer-term lithium and metal-air batteries.

Table 5.3   USABC cost-performance criteria for advanced batteries

Parameter	Mid-term criteria	Long-term criteria
Price	<$150/kWh	<$100/kWh
Range between charges	150 miles (241.5 km)	200 miles (322 km)
Total lifetime range	100 000 miles (161 000 km) (urban)	100 000 miles (161 000 km) (urban)
Calendar life	5 years	7 to 10 years
Specific energy	80–100 Wh/kg	200 Wh/kg
Specific power	150 W/kg	400 W/kg
Climate range	–30 °C to +65 °C	–40 °C to +85 °C
Normal charge	6 h	3 to 6 h
High rate charge	<15 min	<15 min

5.1 Lead-acid

The lead-acid battery was invented in 1859 by Gaston Planté (see Chapter 2), and first used in a vehicle in France in 1881 when G. Trouvé demonstrated its use to power a tricycle capable of travelling at 7 mph. It is interesting to note that today, 120 years later, the lead-acid battery is probably still one of the most widely used electrical storage devices for electric traction applications.

The lead-acid battery, a low-cost version of which is used as the starter battery in all conventional internal combustion engined cars, uses lead and lead-oxide paste plates with a dilute sulphuric acid electrolyte. It has a relatively low energy density of 25–35 Wh/kg in spite of more than 100 years of development, only a modest improvement over the 18 Wh/kg obtainable back in 1909. Most of this improvement is due to weight reduction from the use of lightweight plastic for the battery case material rather than any breakthrough in battery technology.

Power density is relatively good at about 150 W/kg, sufficient to give adequate but not startling acceleration in vehicle use. This figure is reduced as the depth of discharge is increased.

The lead-acid battery is severely affected by low ambient temperatures, and this begins to show up below 10 °C when both energy density and power density are significantly reduced. An electric vehicle using this battery and operated in very cold climates may require auxiliary battery heating and insulation to be fully effective.

As far as battery life is concerned, lead-acid batteries with the type of flat-plate construction used in vehicle starter batteries will survive about one thousand 80 per cent deep discharge cycles, giving an acceptable life of three years. Where battery volume is not a problem – for example, in heavy-duty delivery vehicles – batteries can be constructed with tubular positive electrodes, which gives a significant improvement in life at the expense of some reduction in specific energy. Under well-controlled charge and discharge conditions, lifetimes up to five years can be obtained with this design.

Control of the charging process is particularly important, as it is with most types of battery. It has always been believed that if very rapid charging or overcharging is allowed to take place, severe gassing and shedding of the lead oxide positive grid plate electrode material would occur with rapid deterioration of the battery. Recent work by the Advanced Lead-Acid Battery Consortium (ALABC) has shown, however, that it is possible to charge lead-acid batteries very quickly if the correct charging regime is followed, and this is discussed in Section 5.3 on valve-regulated lead-acid (VRLA) batteries. If the battery is allowed to discharge completely, lifetime is also reduced considerably and for this reason discharge should be limited to a maximum of 80 per cent of full capacity.

Another major failure mode is caused by the formation of corrosion layers at the grid/paste interface of the positive grid. The electrode damage depends on the lead alloy used, and for this reason other additives such as calcium or tin often now replace the antimony originally used in the alloy to improve mechanical strength and ease of manufacture.

Gassing during charging produces hydrogen which requires venting to atmosphere, and in most vehicle installations positive extraction is used to avoid the build-up of a potentially explosive gas mixture.

Lead-acid batteries are, after 100 years of development, better understood than any other battery type, and long experience with them in traction applications means that the problems are well known and can be effectively addressed in the installation design. They also offer a reliable three-year minimum life in continuous traction use, and with their relatively low production cost of about $60 per kWh currently offer the best low-cost option for electric cars.

5.2 Advanced lead-acid

To overcome some of the problems with conventional lead-acid batteries new construction and processing techniques have been developed that have resulted in the advanced lead-acid battery. These techniques have improved the retention of electrode material during high current charge and discharge, decreased the weight of the inactive materials used, optimised the active materials and improved electrolyte mixing.

The result of these actions has been a significant improvement to many of the operating parameters. These include an increase in maximum energy density from 35 Wh/kg to 45 Wh/kg, and a substantial improvement in maximum power density from 150 W/kg to 250 W/kg, made possible by the improved retention of positive plate material during high-discharge operation.

The number of 80 per cent deep discharge cycles before replacement is necessary has been increased from 1 000 to more than 1 500. Battery cost, however, is likely to be about two to three times that of conventional lead-acid when available commercially.

These improvements have been obtained by a number of techniques, including computer analysis and modelling of current distribution. This has led to improved grid structures and designs which have reduced the battery weight and internal resistance, and achieved better retention of the active plate material. There has also been considerable progress in reducing the weight of the inactive materials used for the battery casing, connectors, terminals and separators, by the use of carbon fibre, plastic and plastic lead-alloy materials as well as lead-coated aluminium positive grids and aluminium terminals and connectors.

Developments in active materials and components have included the use of solid reactants with an optimised lead oxide microstructure, a reduction in electrode thickness without any reduction in mechanical strength, and the introduction of methods of electrolyte agitation or circulation to overcome the density stratification common in conventional liquid battery electrolytes.

5.3 Valve-regulated lead-acid (VRLA)

A further development of the advanced lead-acid battery – the VRLA battery – has been the result of a worldwide collaborative effort since 1992 by lead producers, battery manufacturers, component suppliers and related industrial sectors joining together in the Advanced Lead-Acid Battery Consortium (ALABC). The Consortium started from the premise that there was still substantial development potential in the lead-acid battery. This had already been shown to some extent by the improvements made in battery performance by work already done on advanced lead-acid flooded battery technology and described in the previous section, and research undertaken on sealed bipolar lead-acid batteries at AEA at Harwell, UK. In the sealed bipolar technology the electrical resistance of the lead grids and the connectors between them is minimised by the use of a conducting plastic in the cell wall which permits low-loss transmission of current while maintaining the seal between the cells. The cell is sealed to atmosphere and relies on recombination processes in the battery and an increase in the size of the negative electrode to reduce the generation of hydrogen to a minimum. A jelly electrolyte is also used to avoid stratification. An energy density of 50 Wh/kg and a power density of 900 W/kg have been claimed for this technology [1].

The interest of the ALABC when they started was in improving specific energy and therefore range per charge, but it was soon appreciated that this is much less of a problem if it is possible to recharge the battery quickly – 'quickly' in this context being the time that a driver would be prepared to wait at a fast-charging station for a recharge which would allow him to continue a long journey. With this requirement in mind a number of goals were set, as shown below [2]:

- Specific energy of 50 Wh/kg to give a driving range of at least 100 miles
- Specific power of 150 W/kg
- Cost of no more than $150/kWh
- Cycle life of at least 500
- Rapid recharge: 50 per cent in 5 min and 80 per cent in 15 min

Specific power and cost were considered to be already equivalent to or better than the goals selected, so that development effort was concentrated on the other three parameters.

Specific energy has been improved by reduction of the mass of the grid/current collector material in the battery and by an increase in the utilisation of the active masses. Mass reduction has involved the use of lead-calcium alloys with high tin content, which makes it possible to construct thinner, lighter flat-plate and tubular grids without loss of creep strength or reduction in life. Early results with one of the tubular grid designs indicate a specific energy for the system of 42 Wh/kg. Higher active-material utilisation has been pursued by the incorporation of additives, particularly in the positive plate. The most promising of these is porous polypropylene, which is reported to enhance specific energy by over 10 per cent. The combination of this two pronged approach in a single design is expected to result in a battery able to meet the 50 Wh/kg goal [2].

Cycle life has been a major area of development for the ALABC, since early VRLA batteries fell far short of the requirement for a minimum of 500 cycles. The introduction of lead-calcium alloys with a high tin content increased the creep strength, and this improved the cohesion within the positive active grid material and reduced grid expansion to a negligible level. A second factor in reducing cycle life was recognised to be the swelling of the active grid material under deep-cycling operation with a progressive loss of inter-particle contact within the active material and a consequent loss of capacity. To reduce swelling the plate stack is constrained in the perpendicular direction by suitable separators. This has been shown to increase cycle life from around 200 to over 700 [3].

The other major factor in determining cycle life has for many years been recognised as being the charging regime used. This is not too critical for conventional flooded lead-acid batteries, as gassing to atmosphere can accommodate some overcharging. However, since the VRLA battery requires the optimum operation of the internal oxygen cycle, accurate charge management at the top of charge is essential.

To achieve the ALABC rapid recharge goal it was necessary to investigate how VRLA batteries performed under fast battery recharge conditions. Temperature rise in the battery and excessive gassing had always been considered to be limiting factors determining the fastest charge rate that could be used. However, when this was put to the test by fast-charging some 30 commercially available lead-acid batteries, it was shown that all of them could meet the ALABC goal and be recharged to 50 per cent of full charge in 5 min and 80 per cent in 15 min [4]. Not only that but the cycle life was dramatically improved. Two identical batteries were charged, one at 10 A for 10 h, and the other at 250 A for 15 min, and then cycled through repeated discharge and charge cycles. Under these conditions the conventionally charged battery was no longer able to return 80 per cent of its initial capacity after 250 cycles, whereas the fast-charged battery was still healthy after 900 cycles. The coulombic charging efficiency was also improved from between 80 per cent and 90 per cent for the conventionally charged battery to over 95 per cent for the battery that had been fast-charged [5]. Because of minor differences in the cycling regime in the two cases, direct comparison is imperfect, but it is certain that the number of ampere-hours stored and returned was more than three times greater in the fast-charge case.

Fast-charging is done by using computers to control the charging rate while monitoring the battery temperature and internal resistance. The charging characteristics of a typical VLRA battery charged to an interactive pulsed-current/constant voltage algorithm by one-second bursts of high current with a ten-millisecond pause between each pulse to measure internal resistance, are shown in Figure 5.1.

This fast-charging development is undoubtedly a significant break-through in electric vehicle battery technology, giving as it does the possibility of using electric vehicles for long journeys with short visits to fast-charging stations every 100 miles or so. There are, however, significant

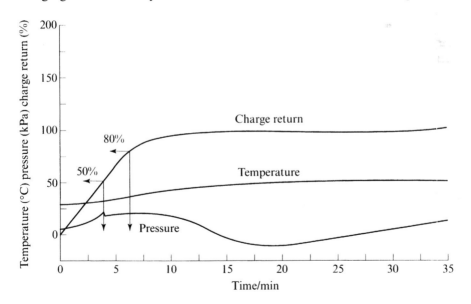

Figure 5.1   Charging characteristics of a VRLA battery recharged to an interactive pulsed-current/constant-voltage algorithm (ALABC)

implications for the infrastructure as far as utilities and the providers of charging stations are concerned. These issues are discussed in more detail in Chapter 7.

5.4 Metal foil lead-acid

Small, metal foil lead-acid batteries have been used in aerospace applications for some time, particularly where high currents for a short period are required. In the last few years Bolder Technologies Corporation has concentrated on developing this technology and combining traditional lead-acid cell chemistry with a much improved mechanical construction. This gives very large plate surface areas combined with unique end connectors that provide very low impedance connections between the plates and the outside world.

The design of a 1.2 Ah metal foil lead-acid cell of 23 mm (0.9 inch) diameter and 72 mm (2.8 inch) length is shown in Figure 5.2. Two layers of coated lead foil and separator are wrapped round a central insulating core so that alternate layers of the lead foil protrude from either end of the coil. The 0.05 mm thick lead foil has 0.08 mm of active lead oxide on each side, with the lead foil layers being separated by 0.2 mm of glass microfibre. This results in flexible plates only 0.25 mm thick, and provides a large surface area which allows more of the active material to be used in the charge/discharge processes. There are also short path lengths to the battery terminals which are cast onto the full length of the exposed ends of the spirally wrapped cell. This construction has resulted in a cell of extremely low impedance, of the order of 0.5 mΩ for the 5 Ah, 2V cell, with the capability of both providing and accepting very large currents when required. The cell design incorporates a resealable vent valve (a Bunsen valve), which permits excess gas to escape if overcharging takes place. Sealing of the battery is straightforward but filling the cells with electrolyte is likely to present some difficulty.

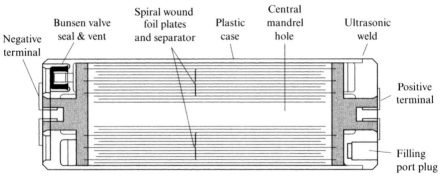

Figure 5.2   Metal foil lead-acid cell (Bolder)

Very high peak power is claimed for this technology, with 900 W/kg up to 70 per cent depth of discharge being reported. Recharge times of less than 15 min are also quoted. It is also claimed that substantial capacity is maintained in low-temperature applications in contrast to some other lead-acid battery types. This makes it potentially very suitable for hybrid vehicles or as a supplementary battery to provide high acceleration for electric vehicles using very high-energy density batteries which have low-power density characteristics (for example the metal-air group).

5.5 Nickel-iron

The nickel-iron battery was invented by Thomas Edison in 1901 to provide electric cars with a longer range than was possible with the lead-acid batteries of the time. Interest in electric cars was high, as they had more than 50 per cent of the then expanding car market (see Chapter 2). With the subsequent virtual disappearance of the electric car, the nickel-iron battery came to be used extensively as a reliable, long-lived but expensive commercial secondary battery in stationary applications.

The battery uses nickel as the positive electrode and iron as the negative, usually with a potassium hydroxide electrolyte. The high exchange current of hydrogen on iron results in considerable gassing, particularly during charging, and also causes corrosion of the iron electrode. This shortens the battery life when not in use by self-discharge. Iron oxides and hydroxides are also transferred to the nickel electrode by ion diffusion. This self-discharge can be quite large, with over 5 per cent of the energy stored in a fully charged battery being lost after only four hours, although the loss rate does decrease as the state of charge reduces. Adding sulphur to the electrode, or lithium, sulphide ions, or hydrazine sulphate to the electrolyte inhibits these processes.

In recent developments energy density has been increased from the 25 Wh/kg originally obtained in stationary batteries, to 50 Wh/kg in prototype development batteries, this improvement being achieved by better utilisation of the active materials and by weight reductions of the inactive battery components. Action that has been taken includes the use of electrodes made of nickel-coated fibres, and of nickel-graphite compounds pressed into nickel wire screens. Sintered reactant powder structures have also been used for both electrodes.

To operate the battery effectively requires the management of its watering and methods of dealing safely with the hydrogen and oxygen produced during discharge. Designs using a single watering point or an auxiliary electrolyte circulating system have also been developed and can give significant performance improvements.

The nickel-iron battery loses capacity at low ambient temperatures, although recent improvements have been reported which allow the battery to perform satisfactorily down to –20 °C. It is capable of a maximum power density of 100 W/kg, which makes it acceptable for adequate acceleration in vehicle use; however, this power density is significantly affected by depth of discharge. The battery has a long lifetime, with up to 2 000 deep discharge cycles, the equivalent of about six years use being possible before replacement. This offsets to some extent the high projected production cost of $150–200 per kWh.

5.6 Nickel-zinc

The rechargeable nickel-zinc battery was invented in 1899 and in the 1920s batteries were constructed but found to have a short cycle life. This was shown to be due to the growth of dendrites on the zinc plate during charging, a condition which short-circuited the battery. At the time no further development was done, but in recent years the high-energy density and power density possible with this battery has increased interest in its use in electric vehicles. Although an energy density of 70 Wh/kg and a power density of 150 W/kg has been obtained, the fundamental problem of dendrite growth (common to all zinc-based batteries), which limits the maximum number of deep discharge cycles to less than 300, has, although improved, not been overcome. It is caused by the high solubility of zinc oxide which is a discharge product of the negative zinc electrode in the alkaline potassium hydroxide electrolyte. This high solubility results in zinc dendrite formation during charging, as well as shape change and densification of the negative electrode on repeated charge/discharge cycles. Also, active material is lost from the negative electrodes by deposition of zinc oxide in the separators.

Lifetime has been improved to 300 cycles of deep discharge by the use of electrochemically impregnated sintered nickel positive electrodes, roll-bonded zinc oxide negative electrodes and special microporous separators which have to be designed to optimise mass transfer while retaining adequate dendrite penetration resistance.

Work has also been done to inhibit dendrite growth by various other methods. These include: using specially shaped electrodes; using additives for both zinc electrode and electrolyte; using vibrating or mechanically wiped zinc electrodes; or making the zinc electrode of slurry held in a matrix. It is also necessary to control charging carefully, starting at a high rate until gas flow as measured by a single pilot cell reaches a specific level, when charging is switched to a much lower rate until gas evolution reaches a second specific level, whereupon charging is automatically terminated.

The battery has a wide temperature tolerance (–39 to +81 °C operating range), a flat discharge characteristic, and a 60 per cent charge reduction after 30 days. However, the sensitivity of this and the other nickel-based batteries to the availability and cost of nickel makes estimates of production prices very uncertain. The best estimate at the moment is $150–200 per kWh.

After a number of attempts to develop practical batteries with acceptable lifetimes in recent years, there has been a substantial drop in research and development on nickel-zinc and other zinc-based batteries. This is unlikely to change unless a significant breakthrough on battery life takes place, although a preliminary assessment of zinc-based batteries being made by the California Air Resources Board (CARB) Battery Technical Advisory Panel, might result in a revival of interest [6].

5.7 Nickel-cadmium

The nickel-cadmium battery cell is constructed with a sintered positive nickel electrode and a sintered or plastic-bonded cadmium negative electrode. These electrodes are interleaved with a highly porous separator which absorbs all the free electrolyte. The battery has an energy density of 50 Wh/kg and a comparatively high power density of 200 W/kg. It is capable of accepting high charge and discharge rates, and this has led to interest from some electric vehicle developers. Nissan have reported 15-min recharge times for batteries with special finned walls designed to dissipate the heat generated during charging. Such high charging rates require careful control and management of battery temperature, voltage and time of charge.

The battery peak power reduces with increased depth of discharge and the number of deep discharge cycles obtainable is reported as being as high as 2 000. This long lifetime offsets, to some extent, the high battery cost. Battery prices, when high-volume production batteries appear, seem likely to be high compared to those for nickel-zinc.

Nickel-cadmium batteries are available in small production quantities from a number of suppliers; as a result they have been used in a number of electric cars. Peugeot-Citroen is probably the largest user in the electric version of the Peugeot 106, of which some thousands have been sold since production started in October 1995. GM and Renault have also considered their use and Pivco Industries, a Norwegian company in which a majority interest was recently acquired by Ford, have produced the Th!nk zero-emission city car using these batteries. This went on sale in Scandinavia in late 1999 and in the USA in 2000, and is claimed to have a range of more than 60 miles at speeds up to 55 mph.

Because of the toxicity of cadmium, the recycling of nickel-cadmium batteries is a complex operation and this has inhibited their use and made other battery types more attractive.

5.8 Nickel-metal hydride

The nickel-metal hydride battery has replaced nickel-cadmium in a number of electric vehicle applications in recent years, since it offers a better performance without the toxicity problems of cadmium.

The GM-Ovonic Battery Company has been the major supplier of nickel-metal hydride battery technology and has developed particular expertise in its application to electric vehicles as a result of being awarded a major contract by the US Advanced Battery Consortium in 1992. The battery has a negative electrode of complex metal hydride alloy compacted onto a conducting substrate, a potassium hydroxide electrolyte, and a positive electrode whose active material is nickel hydroxide. During charging, hydrogen generated by reaction with the cell electrolyte is stored in the alloy as a hydride phase, while nickel hydroxide is oxidised to nickel oxyhydroxide [7].

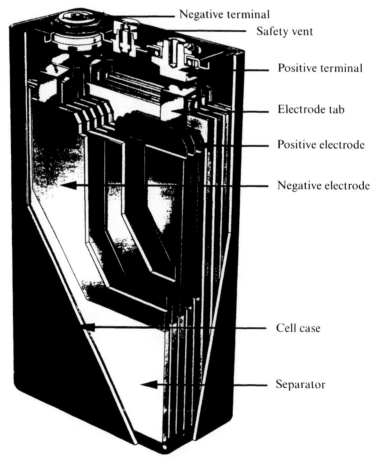

Figure 5.3   *Nickel metal-hydride cell (GM Ovonic)*

The cell construction is shown in Figure 5.3. The cell can tolerate both over-charge and overdischarge because the gas recombination reactions which take place prevent the build-up of excessive pressure inside the cell. This tolerance also simplifies battery management during charge and discharge when many cells are connected in series in an electric vehicle application.

An energy density of greater than 70 Wh/kg is reported combined with a high power density of 200 W/kg. The battery can accept over 600 full charge/discharge cycles to 80 per cent depth of discharge and can be fast recharged to 80 per cent of full capacity in 35 min. Cooling by air or liquid means is being used in the latest batteries to give a reduced package size and to permit faster recharging.

Other manufacturers developing nickel-metal hydride batteries include Varta, SAFT and Matsushita, the last of these has claimed a capability of 1 550 cycles of charge/discharge, giving a potential life of about five years in service. The potential of this battery is shown by its performance in a specially built electric car (Solectria Sunrise) which completed 373 miles on one charge in the 1996 US Tour de Sol, an annual event for advanced vehicles. A GM EV1 completed 245 miles using nickel-metal hydride batteries in the same event.

5.9 Sodium-sulphur

The sodium-sulphur battery was first developed in the Ford Research Laboratories in Dearborn, USA in the 1960s as a new approach to solving the problem of a battery with sufficient energy density and power for an electric vehicle. It is constructed with a positive current collector containing a liquid sulphur positive electrode separated by a beta-alumina separator from a sodium electrolyte in which a metallic negative electrode is embedded (see Figure 5.4). During discharge, sodium ions produced at the negative

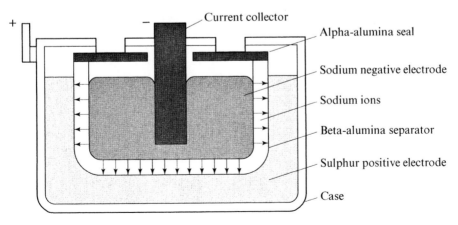

Figure 5.4   Sodium-sulphur battery cell

electrode are transported through the porous separator to the liquid sulphur positive electrode, reducing some of the sulphur to polysulphide ions.

For the battery to operate effectively it is necessary to maintain its temperature at about 350 °C and this requires the use of sophisticated methods of construction for the battery pack, together with auxiliary heating during charging and periods of non-use. It is essential that the battery is not allowed to cool below 200 °C, as the sodium electrolyte freezes solid at this temperature. Reactivation then becomes slow and difficult and the freezing of the sodium can damage the battery because of the mechanical stresses introduced.

The sodium-sulphur battery has a high-energy density of 110 Wh/kg, a high power density of 150 W/kg and is capable of 1 000 cycles of deep discharge in use.

There is concern about the potential problems which could be caused by the leakage of the corrosive materials used in the battery if the vehicle is involved in an accident. Extensive testing has taken place to try to overcome these concerns, including crushing, dropping from heights, exposure to fire and to electrically induced failures, and it is claimed that these batteries survive with no major problems.

The potential problem of excessive heat generation or explosion due to a reaction between sodium and sulphur in the event of an accident is minimised by the cell design. This only allows small amounts of sodium to contact the sulphur even if the solid electrolyte is damaged. In the early 1990s pilot production was in progress at both ABB and RWE-Chloride and it was anticipated that in full production the cost per kWh would be about twice that for lead-acid. Since then ABB stopped production primarily as a result of a small number of fires in 1994 in Ford Ecostar vans using the battery. In December 1995 RWE-Chloride terminated the pilot production programme of its Silent Power subsidiary despite the advanced stage of development which had been reached. This was reported [6] to be due to the reduction in the anticipated EV market caused by the cancellation by California of the zero-emission legislation requirements for 1998–2002. However, the relative success of the next battery to be described may also have had some influence.

5.10 Sodium-nickel chloride

In the sodium-nickel chloride battery (sometimes known as the Zebra or Beta battery), the liquid sulphur positive electrode of the sodium-sulphur battery is replaced by a nickel chloride positive electrode, otherwise the construction is very similar (see Figure 5.5). The sodium salt electrolyte is modified to sodium chloroaluminate which has a lower melting point, at 160 °C, than the sodium electrolyte of the previous battery. It also becomes a powder rather than a solid on freezing, making reactivation easier and

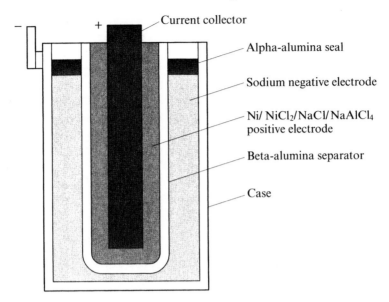

Figure 5.5   Sodium-nickel chloride battery cell

avoiding the mechanical stress problems which occur on freezing in the sodium-sulphur battery.

The battery operates at a slightly lower temperature (300 °C) than sodium-sulphur, has a similar energy density of 100 Wh/kg, and the same maximum power density of 150 W/kg. If damaged it is potentially less dangerous than sodium-sulphur due to the relatively harmless nature of the potential nickel chloride-sodium contact. Another advantage it has over the sodium-sulphur battery is that the cells fail to a short-circuit condition rather than open circuit. This permits the use of single series chains of cells so avoiding the circulating currents present in parallel chains. The cells can also be assembled without the corrosive liquid sodium negative electrode which can be added later.

The Daimler-Benz subsidiary AEG-ZEBRA is now the principal manufacturer having merged in 1987 with the original developers BETA R&D, a subsidiary of the Anglo-American Corporation. The battery is being tested by a number of vehicle manufacturers and there are plans for high-volume production in the future.

5.11 Lithium-iron sulphide

The third high-temperature battery with a potential application in electric vehicles is lithium-iron sulphide. Lithium is particularly suited for use in batteries because of its high electrode potential giving the possibility of very good energy storage.

Unfortunately, lithium has a high reactivity, which limits the type of electrolyte which can be used, and requires stringent safety precautions in cell manufacture and use [7]. Work on the high-temperature lithium alloy/iron sulphide cell has gone on for many years at the Argonne National Laboratory and at SAFT America, with the target being its use in electric vehicles.

The battery has an iron sulphide positive electrode and an aluminium-lithium negative electrode in a non-flammable molten salt electrolyte. The operating temperature is very high at 450 °C with all the additional problems that brings in packaging, insulation and heating. This is offset by the excellent performance with a maximum energy density of 150 Wh/kg and a power density of 300 W/kg. Deep discharge cycle life has been shown to meet the USABC target of 1 000 cycles and cost is expected to be in the same range as the other high-temperature batteries.

The battery does require complex charging controls because it has poor tolerance to overcharging and there are also substantial environmental problems with recycling batteries because of the toxicity of the lithium content.

5.12 Lithium-solid polymer

Lithium has also been used in an entirely different type of battery which makes use of conducting polymers to replace the molten salt electrolyte used in the high-temperature lithium batteries previously described, and in doing so makes possible solid-state batteries of flexible construction. It is possible to produce these batteries in laminated sheet form consisting of a positive electrode of thin aluminium foil coated with vanadium oxide, separated by a foil of solid polymer electrolyte from a negative electrode foil of lithium. This construction permits the sandwich of foils to be cut into appropriate size cells and packaged flexibly as required by the user.

Batteries of this type suitable for electric vehicles have been developed jointly by Hydro Quebec and the 3M Company in conjunction with the USABC, and have shown an energy density of over 150 Wh/kg and a specific power of over 300 W/kg. For best results it is necessary to operate the battery at between 80 °C and 120 °C, although it is possible to operate at reduced power at ambient temperature.

The batteries are rugged and insensitive to shock and vibration damage, can have flexible packaging, and are potentially low-cost. Some 20 V, 119 A modules have been built which can be assembled into a vehicle battery claimed to be capable of giving a range of 150–200 miles in a small car.

5.13 Lithium-ion

In lithium-ion cells the construction, which is shown in Figure 5.6, is similar to that of lithium-solid polymer cells. However, the negative lithium metal plate is replaced by what is known as a 'negative insertion host' such as graphite or tin oxide [8], the lithium being contained within the atomic structure of the negative plate when the cell is charged. During discharge lithium ions travel from the negative host via the organic electrolyte to the manganese, cobalt or nickel oxide positive host, the reverse process occurring during charging. It is said that the lithium ions 'swing' between the cathode and anode, hence the name 'swing' battery which is sometimes used. Cells have an energy density of about 120 Wh/kg and a deep discharge life of 1 000 cycles. Recent information from AEA Technology suggests that this battery can be fast recharged to 80 per cent SOC in less than 1 h.

Development programmes to scale up this technology for electric vehicle batteries are in progress in Japan (Sony), Europe (SAFT and Varta) and the USA (Duracell), funding being from government in Japan, from the European Commission in Europe and through USABC in the USA. Sony have recently announced that they have made a 35 kWh lithium ion battery with a projected life of 3500 deep discharge cycles and an energy density of 120 Wh/kg.

Further development of this battery technology is aimed at replacing the liquid electrolyte with a mechanically strong polymer gel membrane. This plastic lithium-ion (PLI) technology uses a laminate of a carbon-based

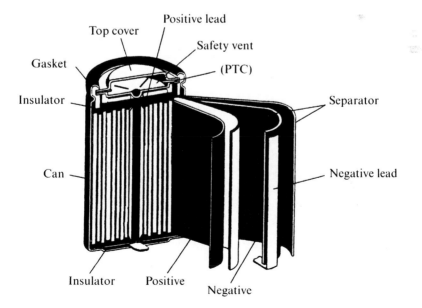

Figure 5.6   Structure of typical lithium-ion cell (from Reference 7)

negative electrode, with the polymer gel membrane electrolyte and a manganese dioxide positive electrode, to produce a cell with an energy density of 100–125 Wh/kg and a life of over 1 000 deep discharge cycles [8]. Currently, PLI batteries are seen as most suitable for use in cellular telephones, portable computers and camcorders, but their properties and potential packaging flexibility also make them of considerable interest for electric vehicle applications. In particular their suitability for low-cost production makes them potentially the best hope for powering economically viable electric vehicles in the next 20 years.

5.14 Aluminium-air and zinc-air

One other class of electric storage batteries exist which have potential for electric vehicle use: metal-air. The metal electrodes which can be used are zinc, aluminium, magnesium and lithium. Of these, zinc and aluminium are likely to be the most useful for this application.

All metal-air battery cells require the use of a thin gas-permeable air cathode and an alkaline water-based electrolyte such as potassium hydroxide. The cell's air cathode absorbs oxygen from the air on discharge and expels oxygen when the cell is being charged. Decomposition of the electrolyte takes place upon prolonged exposure to the air and can be severely affected by varying humidity and carbon dioxide content. This makes it essential for the battery to be sealed against air ingress when in storage. It also makes it essential that in use the air is filtered, humidity-controlled and the CO_2 content reduced by scrubbing.

The rate at which the air-electrolyte exchange takes place determines power density, which is generally low to medium (aluminium-air 30 W/kg; zinc-air 80–140 W/kg), although claimed maximum energy density is very high at 220 Wh/kg for aluminium-air and 200 Wh/kg for zinc-air. The low power density of aluminium-air makes it necessary to use a slave battery in parallel in vehicle operation to provide sufficient power density for adequate acceleration and the ability to absorb power during regenerative braking. It may also be necessary to provide separate electrolyte circulation and conditioning to maintain battery efficiency under operating conditions. This is certainly necessary in the case of the aluminium-air battery where the aluminium electrode is consumed producing hydragilite, which has to be removed and reprocessed. The aluminium electrode also has to be replaced at frequent intervals.

Electric Fuel Ltd in Israel has developed a different approach. Their zinc-air battery has replaceable zinc cassettes which, when discharged, are exchanged for recharged units. The discharged units are then reprocessed at a central charging station. The German Post Office is currently field-testing a fleet of delivery vans with these batteries which are claimed to give a range of over 300 km between charging.

In 1997 a Mercedes-Benz MB410E van using one of these batteries was driven from London to Paris in freezing weather conditions, a distance of 439km (273 miles) on one charge. The combination of a zinc-air battery with a high specific power nickel-cadmium battery capable of handling the high acceleration currents required has been demonstrated in a Nova Transit bus, the combination giving a specific energy of 145 Wh/kg and a specific power of 110 W/kg [9].

It is possible to envisage the use of zinc-air batteries in hybrid vehicles such as long distance trucks and buses, where the zinc-air battery would be used for propulsion in cities. However, as the battery cannot provide load levelling or absorb regeneration energy during braking, some of the lower fuel consumption and emissions reduction advantages of a hybrid would be lost. An alternative to mechanical recharging with its requirement to replace the zinc anode and the potassium hydroxide electrolyte is the recharging of the battery electrically. This can be done by circulating a suspension of zinc particles in the electrolyte through the cell during discharge producing soluble zincate. During recharge the zincate is reduced back to zinc in an electrolysis cell connected to the electrolyte circulation system.

It is interesting to consider if the metal-air battery is properly to be classified as a primary or secondary battery or indeed as a fuel-cell, since it exhibits some of the characteristics of both battery types and of a fuel-cell, depending on the way its charge and discharge cycle is configured and operated.

5.15 Batteries for hybrid vehicles

In the case of batteries for hybrid vehicles the duty cycle and therefore the depth and frequency of battery discharge is very dependent on the hybrid configuration used. In hybrids such as the Honda Insight and the Toyota Prius (see Chapter 10), in which the electric drive is used to assist the main internal combustion engine drive only during acceleration, hill climbing and slow-speed manoeuvring, deep discharge does not happen very often and the battery remains in a good state of charge for most of its life. This should tend to extend its life compared to the battery undergoing frequent deep discharge cycles in a pure electric drive car or in the type of hybrid which is designed to run for extended distances as an electric-only vehicle. On the other hand this 'motor assist' mode of operation used in the Insight and Prius does require significant power from the battery when operating in the 'assist' mode so that high specific power is desirable. A high specific power is also generally an advantage in allowing a correspondingly high recharging rate during regenerative braking.

5.16 Summary – storage batteries

This chapter has covered all the battery types at present being seriously considered for use in electric and hybrid vehicles.

The dramatic improvements in performance demonstrated through the coordinated work of ALABC on valve-regulated lead-acid (VRLA) batteries has made the lead-acid battery once again a contender in the battle to become the battery of choice for the electric vehicle. Its role as a low-cost, well-understood, immediately available power source capable of being fast-charged when required, must make it a strong candidate for the low-cost electric and hybrid cars likely to attract the public in the short term.

Also in the short term nickel-metal hydride seems likely to take over the role of nickel-cadmium as the higher-power, higher-cost option, with its advantages in longer life and safer disposal.

The high-temperature batteries offer significant performance improvement if the vehicle manufacturer is prepared to design systems capable of maintaining battery temperature under the wide range of conditions under which a car can be operated. Of the high-temperature batteries available sodium-nickel chloride seems to have advantages in having a slightly lower operating temperature, better freezing characteristics and failure to a short-circuit condition, so making chains of cells and batteries practical.

In the longer term the metal-air batteries offer considerable improvements in performance, but the special requirements of replacing the metal electrodes and/or circulating the electrolyte need to be shown to be fully practical in the vehicle situation.

The most exciting developments are likely to be in lithium-solid polymer and lithium-ion batteries with their high-energy density, potential for low-cost and flexibility of installation, but volume production still seems to be some years away.

All the batteries that have been described are summarised in Table 5.2. Other methods of storing or generating energy are considered in the next chapter.

References

1   HARBAUGH, D. L.: 'The AEA sealed bipolar lead-acid battery development – Update', ISATA 94, Paper 94EL065, 1994
2   MOSELEY, P. T.: 'The Advanced Lead-Acid Battery Consortium – A worldwide cooperation brings rapid progress', ILZRO/ALABC, 1999
3   MOSELEY, P. T.: *J. Power Sources*, 1998, **73**, p. 122
4   CHANG, T.,G., VALEROITE, E. M., and JOCHIM, D.,M.: *J. Power Sources*, 1994, **48**, p. 163
5   TOMANTSCHGER, K., VALERIETE, E. M., SKLARCHUK, J. *et al.*: 'Laboratory and field evaluations of Optima VRLA batteries utilizing rapid

charging', 13th Annual Battery Conference on Applications and Advances, Long Beach, California, 13–16 January 1998, p. 173

6   MOORE, T.: 'The road ahead for EV batteries', *EPRI Journal*, March/April 1996

7   GIFFORD, P., HELLMAN, V., and ADAMS, J.: 'Development of GM Ovonic nickel metal hydride batteries for electric vehicle and hybrid vehicle applications', 14th EV Symposium & Exhibition (EVS-14), Florida, 1997

8   VINCENT, C. A.: 'Lithium batteries', *IEE Review*, March 1999, pp. 65–68

9   WHARTMAN, J. and BROWN, I.: 'Zinc-air battery-battery hybrid for powering electric scooters and electric buses', 15th International EV Symposium & Exhibition (EVS-15), Brussels, 1998

Chapter 6

Energy sources 2 – Other technologies

There are a number of technologies other than those using chemical batteries which are possible contenders for power storage or generation in electric and hybrid vehicles, and these will be described in this chapter. Some of these technologies are not capable of storing or generating sufficient energy to drive a vehicle over useful distances, but nevertheless some have potential for use as part of the energy storage system to boost acceleration and/or to accept regenerated power during braking.

We will start with descriptions of three methods of storing or generating electrical energy directly, then conclude with two mechanical methods and one thermal method. Of these the fuel-cell shows the most promise at the time of writing, with the flywheel as a longer-term contender.

6.1 The supercapacitor

This storage method makes use of so-called 'supercapacitor' or 'ultracapacitor' technology in which the interface between a conductive electrode and an electrolyte solution forms a boundary layer which can store a charge. This electrode/electrolyte interface has a very small dielectric thickness of a few Angstroms. If this dielectric is combined with a material of high surface area such as carbon black, it produces a low-voltage, high-capacitance, energy storage capacitor in which a theoretical capacitance of one Farad could be provided by a few milligrams of material with an area of about one square centimetre.

Electrochemical double-layer capacitors (EDLCs) of this type have been developed in Japan and Australia using carbon electrodes treated to give an extremely high effective surface area of up to 2 000 square metres per gram, combined with an electrolyte of aqueous acid such as dilute sulphuric acid. Alternatively, a salt solution or an organic electrolyte can be used. The aqueous acid electrolyte gives higher capacitance and power density, while the salt/organic electrolyte gives higher energy density.

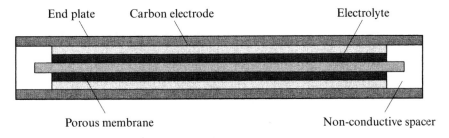

End plate   Carbon electrode   Electrolyte

Porous membrane   Non-conductive spacer

Figure 6.1   Cross-section of a typical cell arrangement for an electrochemical double-layer supercapacitor

In the supercapacitor cell the electrolyte is dispersed and is in intimate contact with the high surface area electrode material so that each cell operates as two capacitors in series. The cell voltage is limited to just over one volt to avoid the voltage gradient across the cell being sufficient to decompose the water in the dilute electrolyte into oxygen and hydrogen. Cells can be connected in series/parallel arrangements to obtain required capacitance values and voltages.

The general arrangement of a typical cell is shown in Figure 6.1.

Recent developments by cap-XX Pty Ltd and CSIRO in Australia have resulted in the production of a supercapacitor of 1 000 F. It is expected that further development work over the next three years will result in a second-generation supercapacitor with an energy density of 15 Wh/kg and a power density of 4 kW/kg, and the ability to provide over 200 A when required [1].

The use of supercapacitors as realistic storage devices for the complete power requirements of an electric vehicle seems to be many years away and considerable development is required if this is ever to be possible. Their high power density does, however, give them an advantage when used in an electric vehicle in combination with either a battery (such as zinc-air) which has a high specific power but a low power density, or in a hybrid drive train as a load-levelling device (see Chapter 10). They can also be recharged very rapidly both from an external power source and by regenerative braking, require no maintenance, do not deteriorate with use and are relatively inexpensive.

6.2 Fuel-cells

Hydrogen is a volatile fuel that can be burnt directly in a conventional internal combustion engine with the emission of water vapour and a very small quantity of the oxides of nitrogen. BMW have been testing concept cars using this technology as a way to meet zero-emission legislation without going to an electric propulsion system. However, this is not the best way to use hydrogen as a fuel source as efficiency is limited to less than 25 per cent by the internal combustion engine Carnot cycle and the losses in the gearbox. If instead hydrogen is used in a fuel-cell to produce electricity which then feeds a highly efficient electric motor driving the wheels directly, an efficiency of more than 40 per cent is obtainable with current technology with potential for up to 60 per cent in the future.

The fuel-cell is a device in which electricity is produced from the reaction of hydrogen with oxygen, the reverse process to that used in the electrolysis of water. Sir William Grove first demonstrated that this was possible in 1836, but it took until the 1960s for fuel-cells to become practical devices for generating electrical energy, the major factor being the development of a suitable electrolyte to control the transfer of hydrogen ions between the two gases. This electrolyte allowed an electrical current to flow in an external circuit from the hydrogen to the oxygen side of the cell, and suppressed the normally explosive reaction when hydrogen and oxygen are brought together and ignited to form water.

Since the 1960s fuel-cells have been used in specialist applications – for example, to generate power in the Gemini and Apollo space missions. Their silent operation and lack of moving parts, combined with their production of potable water as a by-product, made them very suitable as a replacement for more conventional ways of generating power in the special circumstances of a space vehicle.

Various types of fuel-cell have been developed [2]. High temperature (600–1 000 °C) versions using a solid oxide electrolyte (SOFC), or molten carbonate electrolyte (MCFC) and suitable for use in large stationary plant. Intermediate temperature (200 °C) versions using a phosphoric acid electrolyte (PAFC) and suitable for use in small stationary plant; and low temperature (60–230 °C) versions using an alkaline potassium hydroxide electrolyte (AFC) and suitable for use in specialised applications such as powering space craft.

More recent developments in the technology have led to the smaller and lighter proton-exchange membrane (PEM) fuel-cell which operates in the 60–100 °C temperature range. Because of its portability it is particularly suitable for use in automotive and small stationary plant applications. The current leader in developing these fuel-cells is Ballard Power Systems, a Canadian company, which has recently formed an alliance with Daimler-Chrysler and Ford to develop the next generation of fuel-cell-powered vehicles. Other companies are also developing their own PEM fuel-cells, including United Technologies, Siemens, General Motors, Toyota and Mazda, while Johnson-Matthey have been very much involved with the development of catalysts and other components.

The proton-exchange membrane fuel-cell was originally developed by DuPont and combines the anode, cathode and electrolyte in one unit known as the membrane electrode assembly (MEA). Specially designed field flow plates, which have to meet the conflicting requirements of providing good electrical contact and allowing easy passage for the gases concerned,

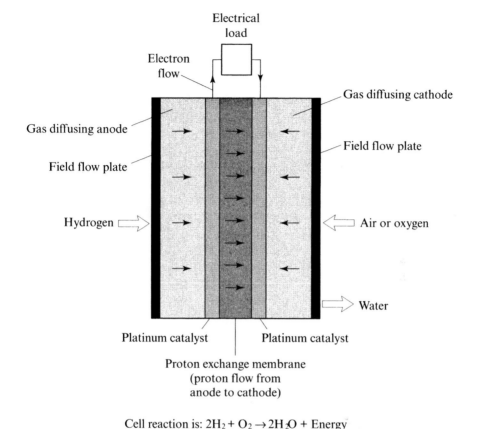

Electrical
load

Electron
flow

Gas diffusing cathode

Gas diffusing anode

Field flow plate

Field flow plate

Hydrogen

Air or oxygen

Water

Platinum catalyst

Platinum catalyst

Proton exchange membrane
(proton flow from
anode to cathode)

Cell reaction is: $2H_2 + O_2 \rightarrow 2H_2O + Energy$

Figure 6.2   Proton-exchange fuel cell

distribute hydrogen to the surface of the anode and air to the surface of the cathode. Both these electrodes are porous, allowing the respective gases to diffuse through them to the proton exchange membrane. In the case of the hydrogen a platinum catalyst bonded to the anode side of the proton-exchange membrane initiates dissociation of the hydrogen into protons and electrons. The electrons then travel through an external circuit while the hydrogen protons travel through to the cathode side of the proton-exchange membrane to which is also bonded a platinum catalyst. This catalyst initiates the combination of the protons with oxygen from the air and the electrons from the external circuit to produce water and heat. The arrangement of the cell components is shown in Figure 6.2.

PEM cells generate electricity at between 0.6 and 0.7 V with a power density of about 1 W/sq.cm [3], equivalent to about 120 W/kg for a cell stack. Fuel-cell efficiencies vary between 40 per cent and 65 per cent. Specific energy is dependent on the quantity of hydrogen available to feed the fuel-cell. One disadvantage that fuel-cells have when used in a vehicle is the large number of 0.7 V modules required to obtain a sensible working voltage (200–300 V) and therefore the large number of electrical and gas connections also required. There is also a potential problem in cold climates with the handling of both the water produced by the fuel-cell and that used if a cooling system is required. In these circumstances freezing may occur in standing vehicles with potential damage to the fuel-cell. Most PEM fuel-cells suitable for automotive use also require air compression to about 3 atmos or higher [4] to improve cell power density, and this causes some loss of efficiency.

When fuel-cells are used in a vehicle it is necessary to provide them with a supply of clean hydrogen. The various methods of doing this are described in Chapter 11 together with a discussion of the infrastructure issues that arise in the practical application of this technology in an electric car to be used by the general public.

6.3 Solar cells

Solar cells using silicon in amorphous or crystalline form have been used to power electric cars specially designed for extremely low power consumption and low-speed operation with a single lightweight driver. Cars of this very specialised type compete annually in the USA Sunrayce from Washington to Orlando, Florida, and in the Darwin–Adelaide trans-Australia international solar car race. These cars are not practical vehicles that could be used for everyday transport, being designed for minimum rolling and wind resistance at the low speeds normally used. The power consumption of these specialised vehicles is extremely low, of the order of 25 Wh/km (40 Wh/mile). This should be compared to the 160 Wh/km (260 Wh/mile) required by the best of the purpose-built practical electric cars such as the GM EV1.

With this very low power consumption and a large array of solar cells covering the whole vehicle, it is possible in a sunny climate to provide sufficient charge to keep a vehicle of this type running. Both the USA and the trans-Australia races have been useful test beds for new solar cell designs, as has the space industry.

When the first solar cells appeared, having been developed by AT&T Bell Laboratories in 1954, they had conversion efficiencies of 6 per cent. Solar cells now available have efficiencies of 20–30 per cent and a power density of 150 W/m^2 from normal sunlight. However, power density is drastically reduced in cloudy conditions or where the solar-cell array is shaded or tilted relative to the direction of the sun, and of course is zero between sunset and sunrise. Since all practical vehicles need to operate under this range of conditions, battery storage of the energy produced when the cells are in sunlight is essential if a vehicle fitted with solar cells is to keep running. Even allowing for efficient storage of the energy generated the average power density over 24 h in a very sunny part of the world is probably not much more than

$30 \, W/m^2$. Given the restricted maximum area of $2–3 \, m^2$ of solar cells it is possible to mount on a vehicle, it is clear that the energy generated is only sufficient to propel a vehicle of the special type used in the Darwin–Adelaide race, not a vehicle in the GM EV1 category.

Solar cells can be used very effectively for powering a fan to ventilate the interior of a vehicle parked in sunlight, thus reducing air-conditioning load at start up. This could be particularly important in electric vehicles where the air-conditioning load can consume up to 50 per cent of the energy needed to drive the vehicle. Another use is for providing a trickle charge to the starter battery of a conventional internal combustion engined vehicle when it is parked for a long period.

One application of solar cells that is of potential interest for an electric vehicle, is the conversion of thermal radiation from a burning fuel directly to electricity. This technology, known as 'thermophotovoltaics' (TPV), was first proposed at MIT in 1954, when it was realised that the power output of a solar cell could be considerably improved by using the infrared radiation from a radiant source instead of sunlight. Using thermal radiation from a source at $1\,500 \, °K$, prototype TPV cells have been shown to yield a power density more than 60 times that of a solar cell in sunlight [5]. The application of this technology to electric vehicles has been demonstrated in a prototype car developed at the University of Western Washington that is partially powered by TPV convertors run on propane. Any fuel that can be burnt efficiently can be used and an overall efficiency for a practical system of up to 30 per cent is obtainable.

6.4 The flywheel

The storage of energy in a flywheel is an ancient technology. Thousands of years ago flywheels were used to produce smooth rotation of the potter's wheel from the pulse input provided by the human foot. In more recent times flywheels have been used to smooth out the rotation of steam engines and internal combustion engines. In both cases the pulsed input energy is stored as rotational kinetic energy in the flywheel according to the equation: energy stored $= kr^2m\omega^2$. In this equation r is the radius of the disc, m is its mass, ω is its angular velocity, and k is a constant depending on the geometry or shape of the flywheel disc (k is 0.5 for a thin rim disc; 0.6 for a flat disc; and 0.8 for a truncated conical disc designed to obtain equal stress throughout the flywheel and therefore maximum energy storage density). This equation shows that if large amounts of energy are to be stored, it is essential that the flywheel operates at a high angular velocity, since energy stored increases proportionately with mass, but as the square of angular velocity.

Where a flywheel is used to smooth out the rotation of a conventional internal combustion engine in a vehicle, the amount of energy stored is quite small and is limited by the need for the engine to be able to accelerate reasonably quickly when required. If, however, it is to be used to store significant amounts of energy, it is necessary to optimise the design of the flywheel itself and the method used to transfer energy in and out. The ultimate limit on the amount of kinetic energy that can be stored in a particular flywheel is determined by the tensile strength and density of the material used in its construction; these factors are related by the following equation: energy density = $E/m = k (\sigma/\rho)$ where σ is the tensile strength of the flywheel material, ρ is its mass density, and k is the constant used in the earlier equation for energy stored, and is determined by the geometry of the flywheel. This equation indicates that if a high-energy density is to be obtained, it is essential to make the flywheel from light material with high tensile strength.

Until recently flywheels were made of steel. An optimised steel flywheel would have a theoretical energy density of less than 50 Wh/kg. However, a flywheel made of a composite material such as carbon fibre or kevlar can be shown to have a theoretical energy density of about 200 Wh/kg. These figures indicate how much energy can theoretically be stored as kinetic energy in a flywheel of given mass and material before the limiting tensile stress is reached. The practical energy density will be reduced to less than 50 Wh/kg because the mass of the flywheel mountings and containment shielding has to be taken into account and a safety factor included to avoid any risk of mechanical failure.

The area where the flywheel is at its best is in being able to generate or absorb very high amounts of power for short periods at an efficiency as high as 98 per cent; this compares with a turnaround efficiency of 75–80 per cent for chemical batteries. In vehicle use this means high acceleration and effective absorption of power generated by regenerative braking. Power densities up to 2 000 W/kg are possible without exceeding safe working stresses. The flywheel also has one other major advantage over chemical methods of energy storage; its life is indefinite and its level of performance constant irrespective of the number of deep discharge cycles undergone, providing the elastic limit of the materials used is not exceeded and catastrophic mechanical failure is avoided. This can be a significant factor in determining the economics of a flywheel power source versus a chemical battery, since the chemical battery shows a progressively deteriorating performance with age and requires replacement at a cost of a few thousand dollars every three to five years.

Although the maximum energy density of a flywheel is independent of its speed, for a particular energy density the flywheel diameter is inversely proportional to its speed of rotation. There is therefore an advantage in using a small-diameter, high-speed flywheel since this will have a reduced gyroscopic effect in vehicle use, and will also reduce the angular momentum transferred to the vehicle if the flywheel were to be suddenly stopped in an accident. The general trend among developers of this technology is to go for modules consisting of pairs of small, high-speed composite contra-rotating flywheels (to cancel gyroscopic effects), capable of storing 1 to 2 kWh of energy each at maximum rotational speeds up to 100 000 rpm. These modules can then be assembled for use in both electric and hybrid vehicles to make up the power storage requirement of 20 to 40 kWh for a pure electric and 4 to 8 kWh for a hybrid. It has been suggested that it would be possible to double rotational speeds to 200 000 rpm so obtaining a four-fold increase in power stored and giving a small car a range of 300 miles.

A typical design for a carbon-fibre flywheel is shown in Figure 6.3. The concentric carbon-fibre shells of the flywheel rotor are surrounded by a substantial containment shield to avoid parts of the flywheel breaking out in an accident. Near the centre of the flywheel and embedded axially in the carbon-fibre matrix are permanent magnets which serve both as rotor magnets for a motor generator to transfer energy in and out of the flywheel, and as one half of the radial magnetic bearing which provides support for the rotor. The static inner shaft then carries both the motor generator stator coil and the radial magnetic bearing coil. The use of a magnetic bearing and the elimination of air friction by the evacuation to a high vacuum of all spaces inside the containment shell is effective in reducing frictional losses to a very low level and gives a flywheel that will continue to rotate for a very long period when not loaded. To increase the energy that can be stored, the concentric carbon-fibre shells making up the flywheel may be put under radial compression. This causes the centrifugal loading on the flywheel as it

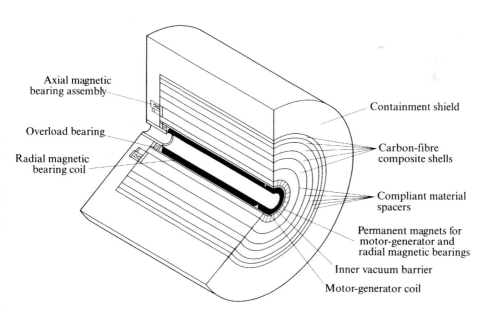

Figure 6.3   Carbon-fibre flywheel with magnetic bearings and permanent magnet motor-generator

speeds up to first have to neutralise the radial compression before going into tension, so making it possible to store more energy than would otherwise be possible.

There are important safety issues that have to be considered if flywheels, or electromechanical batteries as they are sometimes called, are to be used in an electric vehicle. Because of the large amount of energy that may be stored and the catastrophic effect if it were suddenly released, containment must be effective even when a crash occurs. The use of a composite flywheel helps considerably in meeting this requirement as if one of these flywheels were to burst it would disintegrate into small fragments that behave like a fluid instead of the large pieces which would be ejected from a bursting steel flywheel. Although the actual disintegration only absorbs part of the energy, the fluid-like behaviour of the fragments does mean that that the remaining energy is more equally distributed around the inside of the containment vessel and can be better absorbed by being distributed over a large area. It is also possible to design the inner surface of the containment vessel to direct the fragments in the most advantageous direction, or even to allow the containment vessel to rotate relative to an outer stationary shell to absorb the angular momentum.

Most of the development of flywheel energy storage is currently being undertaken in the USA and Canada, in particular at American Flywheel Systems, Flywheel Energy Systems Inc., Lawrence Livermore National Laboratory, Oak Ridge National Laboratory, Penn State University and University of Texas. It seems probable that we are still some years away from a viable commercial product using this technology.

6.5 The hydraulic accumulator

The hydraulic accumulator operates in a similar way to a flywheel in that it can store energy during braking, low-speed or idle operation in a hybrid vehicle (see Chapter 10). It can then feed this energy back to the drive system when high power is required during acceleration. The accumulator stores energy by pumping oil at high pressure into a pressure vessel containing a highly deformable membrane that separates the oil from a compressible gas. The oil is pumped in using a combined pump/hydraulic motor, the hydraulic motor then allowing the stored energy to be withdrawn as required. The amount of energy that can be stored is not sufficient to power a vehicle for any significant distance. However, the hydraulic accumulator is a possible replacement for a flywheel or high-discharge battery in a vehicle where the long-term energy source is not capable of providing the power for acceleration or absorbing the power generated in regenerative braking.

6.6 Compressed-air storage

Although not an electrical method of energy storage, compressed-air storage is of interest both because it could perform a similar function to a hydraulic accumulator in a hybrid system and because it also appears to be capable of providing sufficient power to propel a car over a significant distance.

In 1994 it was reported that compressed-air storage had been used to power an experimental car [6]. Researchers in Joplin, Missouri used three air cylinders filled with air compressed to 200 times atmospheric pressure to power two compressed-air motors, each driving one of the front wheels. Each motor has two double-acting cylinders so that the compressed air can be used at both high pressure and low pressure to obtain maximum efficiency. A top speed of 60 kph was claimed for the 1988 Chevrolet Sprint used, and one full charge of the compressed air cylinders, which takes only four minutes, was reported as lasting two and a half hours. The range implied by these figures, although not claimed in the report, is at least 100 km, which makes the technology, known as the Pneumacon system, of more than passing interest.

6.7 Thermal energy storage

Thermal energy storage can include both high-temperature thermochemical and Latent-heat methods, using eutectic salts such as barium hydroxide, which change state when taking in and giving out energy. The use of such an energy storage method is only relevant for regeneration and load-levelling in vehicles which make use of specialist prime movers such as the Stirling engine and the steam engine, and are of no interest for electric vehicles unless they are hybridised with such engines. However, it is possible, using these methods, to store engine heat in internal combustion-engined vehicles and make this heat available for the rapid heating and demisting of the passenger compartment, where otherwise direct electrical or hydrocarbon fueled heaters may be necessary. Thermal energy storage could also be used to store heat generated during charging to provide rapid heating in electric vehicles.

6.8 Summary – non-battery energy sources

Of the non-battery methods of providing power for an electric car, only fuel-cell generation and flywheel storage provide sufficient power to operate an electric car over a commercially useful range. Recent developments in fuel-cells make them a more likely power source for the shorter term than flywheels. The proposals for fuel-cells which can run directly on methanol or gasoline, rather than as at present having to use reformers to convert these fuels to hydrogen, make fuel-cells a potentially more attractive short-term proposition than hitherto. If this development is successful we could see significant numbers of electric vehicles using this technology within ten years. The successful development of an infrastructure for supplying clean hydrogen direct to vehicles could, however, change the emphasis to the use of hydrogen storage on the vehicle. These issues are discussed in Chapter 11.

Flywheels still have a large safety hurdle to overcome. Before they can be accepted in production vehicles it will have to be shown conclusively that the flywheel containment system can cope with all the accidental failure mechanisms which can occur in a vehicle in use on the public roads, and that the gyroscopic effects both in normal use and in a catastrophic failure do not increase the danger to the user.

The other energy sources described in this chapter cannot, with perhaps the exception of compressed air storage, store sufficient energy to propel a vehicle over a distance great enough to be of interest. However, because of their generally high power density, they can be very useful when hybridised with other energy storage methods such as chemical batteries, which have a high-energy density but low power density (for example metal-air batteries); or with other primary power sources such as internal combustion engines. The use of these technologies in hybrid vehicles is discussed in more detail in Chapter 10.

References

1   VASSALO, T. and PAUL, G.: 'High density energy storage – supercapacitors', *Interface*, CSIRO Energy Technology, North Ryde, Australia, July 1997, p. 3
2   LARMINIE, J.: 'Fuel cells come down to earth', *IEE Review*, May 1996, pp. 106–109
3   HOOGERS, G.: 'Fuel cells: power for the future', *Physics World*, August 1998, pp. 31–36
4   WIENS, B.: 'The future of fuel cells', June 2000, Internet www.benwiens.com
5   COUTTS, T. and FITZGERALD, M.: 'Thermophotovoltaics: the potential for power', *Physics World*, August 1998, pp. 49–52
6   BEARD, J.: 'Spirit of Joplin blows into town', *New Scientist*, 6 August 1994, p. 21

Chapter 7

Charging

The charging and recharging of electric vehicle batteries is a critical part of the energy cycle in an electric vehicle. Although less attention is usually paid to charging than to the vehicle batteries and the electric motors that they drive, the availability of efficient, and where appropriate fast, recharging of batteries is a vital factor in deciding whether a particular electric vehicle is of practical use for regular transportation.

Most electric cars will undoubtedly be charged at home from the domestic power supply, and if the vehicle is used for the short urban journeys usually envisaged for electric cars, this slow, eight-hour charge overnight from a conventional power point providing 3–4 kW is probably adequate. However, if electric cars are ever to widely replace those using the internal combustion engine, then much faster and more convenient methods of charging will also be required. This chapter will describe both basic and more sophisticated charging methods, their usage with different battery types and their implications for the infrastructure. The requirements for recharging energy storage systems other than those using chemical batteries will also be described.

7.1 Early systems

In the early years of the 20th century when the battle for supremacy between the electric car and the internal combustion engined car was at its height (see Chapter 2), charging systems were simple in design and construction. Where an AC supply was available a transformer was used to provide a voltage suitable for the battery in use – usually lead-acid or nickel-iron – this then being full-wave rectified and the resulting DC smoothed by a capacitor and fed to the battery through a rheostat which could be adjusted to provide a suitable current. A ballast resistor was also included to limit the charging current to a suitable maximum level. The basic circuit of this simple charger is shown in Figure 7.1. A typical

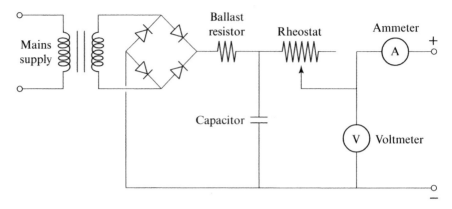

Figure 7.1   Early charger circuit for use with AC supply

installation of this type from 1912 is shown in Figure 7.2. To make charging possible when only a DC supply was available it was necessary to either ensure that the vehicle battery voltage was lower than the supply voltage or to provide DC-to-DC step-up conversion by the only method then possible: a motor-generator. An example of this form of charging is shown in Figure 7.3 where an early electric car, probably of 1912–14 vintage, is shown being charged through a motor-generator set mounted in a vertical

Figure 7.2   Charge installation from 1912

Figure 7.3 Early electric car being charged from a low voltage DC supply by a DC motor/DC generator set

unit near the main supply switch. With the DC supply produced by the motor-generator, only a rheostat and limiting resistor together with an ammeter and voltmeter was then required to complete a charging installation.

7.2 Charging techniques for modern lead-acid batteries

With the now universal use of AC domestic and industrial supply systems, the general principles of the early AC charger design described above have continued in widespread use to the present day. The current into the battery is, however, usually controlled electronically by the continuous feedback of critical battery parameters, and the limiting ballast resistor is replaced by an inductive reactance, either in the form of a choke in series with the transformer or as leakage reactance built into the transformer.

The characteristics of this charging circuit are shown in Figure 7.4. It will be observed that in the initial phase of charging a lead-acid battery the voltage rises steadily from 2.1 to 2.35 V per cell, at which point the battery is about 80 per cent charged and hydrogen and oxygen evolution (gassing) at the electrodes commences. After this the voltage rises rapidly to a maximum value of about 2.45 V and as a consequence the charging current drops, its reduction being determined by the internal impedance of the charger.

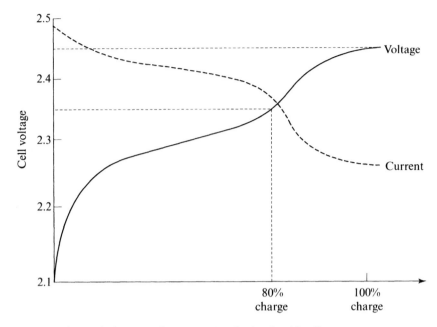

Figure 7.4   'Taper' charging characteristics for lead-acid cell

Because of this tapering characteristic this type of charger is known as a 'taper' charger. It should be noted that gassing has a useful function in promoting mixing of the electrolyte and the avoidance of stratification which can have the effect of reducing the apparent capacity of the battery to accept charge.

In more sophisticated chargers for lead-acid traction batteries and particularly for valve-regulated lead-acid traction batteries in which the hydrogen and oxygen evolution within the battery cannot under normal conditions vent to atmosphere, a more complex charging regime is used. The charging schedule is shown in Figure 7.5 and is a combination of a constant-current and a constant-voltage step in which the current and voltage levels used are determined by the battery temperature as sensed by a temperature probe. This temperature probe is mounted on the base of the battery and insulated from external temperature effects. In the initial charging phase A, the maximum constant current that the battery will accept or that the charger can provide is maintained until gassing starts, without raising the battery temperature above an acceptable level. The voltage is then maintained at a constant level during charging phase B while the charging current decreases until at time T_1 it has reached a level of about 45 per cent of the maximum, at this point the remaining time T_2 required to complete charging phase B can be determined. This procedure has the advantage that it makes allowance for the state of charge of the battery at the start of charging, since if a nearly fully charged battery is connected to the charger the current will

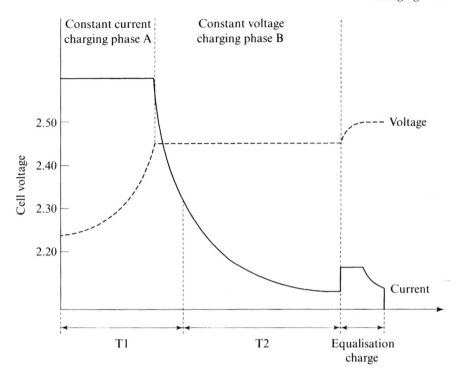

Figure 7.5   Charging schedule for valve-regulated lead-acid batteries with gelled electrolyte

decay very rapidly due to the early onset of gassing and a short charging time will result.

Because the individual cells in a lead-acid battery differ in their acceptance of charge during the constant-voltage gassing phase (due primarily to differences in their internal oxygen cycles), it is necessary to provide a further shorter period of charge at a higher voltage to equalise the individual cell charge status, and this is known as the 'equalisation' charge. This equalisation charge follows directly after the main charge and normally lasts for one hour with the current limited to less than 10 per cent of the maximum charge acceptance rate of the battery. To obtain this level of current in a fully charged lead-acid battery cell requires the application of about 2.5 V, but it is important that this equalisation charge is not prolonged or overcharging can result in damage to the battery by accelerated grid corrosion or thermal runaway. Thermal runaway occurs when heat generation within the battery increases more rapidly than it can be dissipated by the usual processes of convection and conduction. It is caused usually by an overvoltage, which results in an increase in the rate of reaction in the oxygen cycle, which increases heat generation and water loss, which then in turn further increases reaction rate. This cycle can continue until the battery is destroyed. Valve-regulated lead-acid batteries with a gelled electrolyte are less prone to this problem than those using a conventional electrolyte. Thermal runaway is not normally a problem in vented batteries.

Recent work under the auspices of the Advanced Lead-Acid Battery Consortium (ALABC), has demonstrated the practicality of fast-charging VRLA batteries [1], in particular by the use of partial state of charge (PSOC) operation, in which the battery is charged at a high rate from 20 per cent SOC (state of charge) to the 80 per cent SOC level at which gassing starts, then discharged under normal operating conditions until the SOC has fallen to 20 per cent again. This PSOC operation is combined with the use of a gelled electrolyte to avoid the charge stratification that occurs in liquid electrolytes under part-charge conditions. This method of charging has been shown to considerably improve the number of deep discharge cycles the battery can survive. It is necessary to provide a full charge every three or four PSOC charge cycles and an equalisation charge every 12 to 15 PSOC charge cycles.

In the tests reported in [1], a 33 Ah vehicle battery was charged three or four times per day at 165 A on a 20 Ah PSOC cycle, returned to 100 per cent SOC at the end of each day, and given an equalising charge every fourth day. The vehicle was driven 16 846 miles with the battery modules receiving 697 high charge rate cycles and 738 cycles in total and delivering 15 258 Ah during the test period. The high current PSOC charge allowed recharging to 80 per cent SOC to take place in about 8 min, with full charging to 100 per cent SOC at the end of each day taking about 15 min. This fast-charging regime permitted the vehicle to continue in use throughout the day, making it a practical proposition for commercial use, and demonstrating that the use of lead-acid batteries is a viable, low-cost option for the electric vehicles of the future.

Fast-charging of the type described above is normally done under computer control with feedback of the battery temperature and internal resistance. Charging is then done with an interactive pulsed current/constant voltage algorithm using typically one-second bursts of high current with a ten-millisecond pause between each pulse to measure internal resistance. Pulsed charging is particularly well suited to computer-control methods and seems likely to be widely used in future electric vehicle charging systems.

Apart from the somewhat idealised recharging schedules just described, it is frequently necessary to recharge electric vehicles wherever the opportunity arises if a useful range in mixed operation is to be obtained. This form of recharging is known as 'boost charging', 'opportunity charging' or in parts of Europe 'biberonage'. Charging is done at as high a rate as is permitted for the battery concerned or is available from the charging point used, with the charge usually being limited to the constant-current non-gassing charging phase. When operated under these conditions the battery should receive a full charge at least once a day, although this is less important with valve-regulated lead-acid batteries using a gelled electrolyte since the stratification induced by this partial charging in a liquid electrolyte is avoided.

7.3 Charging techniques for nickel-based batteries

There are significant differences between the operation of those nickel-based batteries (nickel-iron, nickel-zinc, nickel-cadmium and nickel-metal hydride) which are suitable for electric vehicle use, and of the lead-acid battery systems discussed in the previous section. In particular, prolonged overcharging is less critical because the internal oxygen cycle, which absorbs the excess charging current, is moderated by the high oxygen reduction rate at the negative electrode. This makes it possible for nickel-based batteries to be overcharged at much higher rates than lead-acid batteries without losing water. Nickel-cadmium batteries have a high initial charge acceptance up to 50 per cent of full capacity and high charging rates are obtained at fairly low cell voltages; above 50 per cent SOC the internal oxygen cycle takes more current and it is necessary to raise the voltage to maintain the charge rate. If high charge rates are to be achieved, it is necessary to control the charging current to avoid excessive temperature and pressure within the cell, and this is best done by feedback of cell temperature and open-circuit cell voltage. Figure 7.6 shows the cell voltage and temperature during uncontrolled fast-charging and illustrates the reduction in cell voltage near full charge caused by the increase in cell temperature as a result of the acceleration of the internal oxygen cycle.

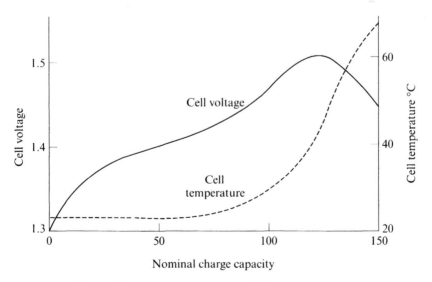

Figure 7.6   Cell voltage and temperature in nickel-cadmium battery during uncontrolled fast-charging

Because of the negative slope of the voltage curve thermal runaway can take place if the charging current is not reduced.

For very rapid charging of nickel-cadmium batteries it is essential to use pulse charging to enable the open-circuit battery voltage to be measured during the period charging is interrupted. Together with temperature feedback and electronic control it is then possible to operate a constant-current initial period of high-rate charge until a pre-set open-circuit voltage is reached, followed by a period in which the charging current is progressively reduced to maintain the battery temperature sensibly constant. The characteristic for this type of controlled charging is shown in Figure 7.7. It is possible to recharge nickel-cadmium batteries in 15 min using these methods.

The charging characteristics of nickel-metal hydride batteries are similar in almost all respects to those of nickel-cadmium, and replacement of one type by the other in a system should present no problems.

One characteristic of nickel-cadmium batteries that should be considered and is of some importance if batteries are subject to a number of slow partial charge-discharge cycles without full discharge being completed, is the 'memory effect'. Under these charge-discharge conditions intermetallic nickel-cadmium compounds may be formed and cadmium crystals can grow on the cadmium electrode, reducing its surface area and causing an increase in the voltage drop during discharge that results in an apparent reduced capacity. The condition can be quickly corrected by complete discharge followed by full recharge. Generally this condition is not a problem in electric vehicle batteries as relatively fast charge and deep discharge is normal, although there may be a problem in hybrids where the SOC may remain in

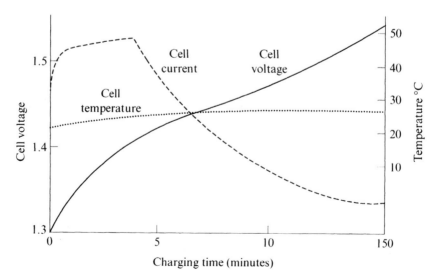

Figure 7.7   Controlled rapid pulsed current charging of a nickel-cadmium battery

the middle range for a considerable period. Chargers can also be designed to complete discharge of a battery before starting a charge sequence.

7.4 Charging techniques for non-aqueous batteries

Non-aqueous batteries (sodium-sulphur, sodium-nickel chloride, lithium-iron sulphide, lithium-solid polymer and lithium-ion) are less able to accept overcharging than the aqueous batteries described in the previous two sections. For this reason charging is usually to a predetermined fixed maximum voltage using either constant-current or taper-charging techniques. No information is available at the time of writing on fast-charging techniques for this class of battery.

7.5 Battery state-of-charge measurement

Any measurement of the SOC of a battery should ideally correspond exactly with the amount of charge available for withdrawal from the battery under normal operating conditions. In a conventional vehicle driven by an internal combustion engine it is comparatively straightforward to determine the quantity of fuel remaining in the tank, by measuring the depth of fuel remaining. In the case of an electric vehicle there is no simple way of measuring the SOC of a complex battery system, since this system may require many cells in series and parallel to obtain the 200+ V and the high current required for operation. The obvious parameter, which is easily measured, is voltage; generally, however, as shown in Figure 7.8, batteries have a relatively flat voltage-discharge curve under moderate load until close to the complete discharge point. A flat discharge curve is a highly desirable characteristic from an operational point of view, since it is important that voltage does not fall much over the normal operating range. If it does, it seriously affects the ability of the system to maintain power to the drive wheels. Voltage is therefore a very unsatisfactory parameter to use to indicate SOC although it is widely used because of its simplicity.

In batteries using an aqueous electrolyte, particularly vented lead-acid, measurement of electrolyte specific gravity is a comparatively accurate way of determining SOC. It is, however, very difficult to continuously measure individually the specific gravity of the large number of cells used in an electric vehicle battery, even if the variations in specific gravity due to electrolyte stratification could be overcome and the problems of access to the electrolyte resolved. One proposal that has been made is to measure specific gravity by sensing changes in the optical properties of the electrolyte in a lead-acid battery. This is done by inserting an optical fibre in the electrolyte and measuring the light transmitted to a second adjacent fibre, or alternatively measuring the change in light transmitted down an unclad optical fibre

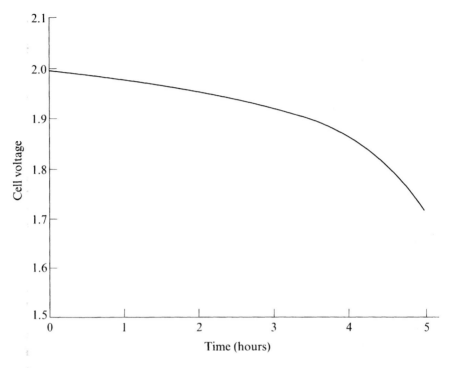

Figure 7.8   Typical discharge curve of a lead-acid battery under moderate load

immersed in the electrolyte [2]. It is also suggested in [2] that SOC in a lithium-ion battery could be measured by introducing an optical fibre into the anode to monitor absorption of light when lithium ions are introduced.

It is generally considered that there are three possible practical methods of measuring SOC in a battery containing a large number of cells. The first of these is by the measurement of the response of the battery to an applied or removed load. The second is by the measurement of the battery impedance. The final method is by the measurement and continuous integration of the power taken from the battery during discharge and the power returned to it during charging, combined with a method of allowing for losses and establishing the true SOC at as frequent intervals as possible. This is usually known as the coulomb method.

An example of the first method is described in [3] and proposes a method of measuring the SOC of lithium batteries by recording the battery temperature and the time profile of voltage recovery after a fixed load is removed and comparing these values with reference tables for these characteristics. The second method in which SOC is derived from battery impedance measured in various ways has been proposed by a number of inventors. It has been proposed that the change in battery capacitance [4], the change in complex phase angle [5], and the change in battery impedance at two different frequencies [6] be measured and compared with a look-up table or a stored battery model. However, the most promising approach in the author's opinion, is the final one, the so-called 'coulomb method', particularly when this is combined with one or more of the other two methods.

A system of this type is described in [7], and although intended for the measurement of the SOC of conventional automotive 'starter' batteries, it has potential for an acceptably accurate measurement of the SOC of electric-vehicle batteries. The method may be considered to be a combination of the coulomb method with a variant of the response of the battery to an applied and removed load.

In this system the battery charge input and output are time integrated and compared with a battery model stored in a microprocessor. This model includes an allowance both for battery temperature and for efficiency losses during charging. The dynamic SOC is then given by the balance of charge and discharge compared to the original static SOC of the battery. The overall static SOC of the battery is determined by observing the open-circuit voltage of the battery during periods of zero charge or discharge. Under these conditions the open-circuit voltage is a reasonably accurate measure of the electrolyte density in a lead-acid battery and therefore its true SOC providing the battery remains in the zero-charge and discharge condition for a sufficient period. That this period is quite long is shown by the voltage-stabilisation curves from [7] shown in Figure 7.9 (a) and (b), which can be seen to require up to three hours to reach a stable level after a period of discharge, and ten to twelve hours after a period of charging. Clearly it is not practical for an electric vehicle to stand idle for this period of time frequently enough to ensure accurate resetting of the SOC measuring system. However, using the stored battery model it is possible to predict the eventual voltage stabilisation level from the rate of voltage change within the first few minutes after charge or discharge ceases. From this prediction static SOC values can then be determined accurately enough to provide a substantially more accurate dynamic SOC measuring system than has been available up to the present. It would be possible to incorporate this short period of zero charge and discharge into a slow-charge algorithm when overnight charging is being undertaken. The system can also indicate deterioration of the battery with age as the open-circuit voltage decreases under these conditions.

7.6 Battery management

Active battery management in electric vehicles requires not only effective and accurate continuous measurement and control of state of charge, voltage, temperature, and charge and discharge rates for the whole battery system, but also monitoring of the voltage and temperature of individual battery cells for faults such as low voltage, overheating, high resistance, or open or short circuit. To

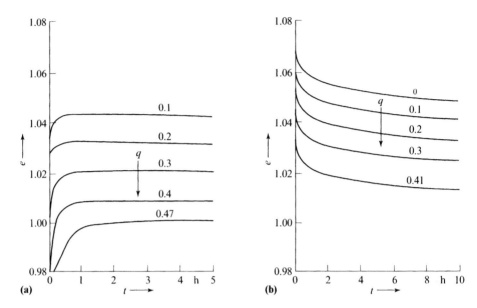

Figure 7.9   Stabilisation of the open-circuit voltage of a lead-acid battery after (a) a period of discharge and (b) a period of charge (from Reference 2) where e is the normalised rated voltage for a given battery (for example for a 12 V battery 1.00 is 12 V), q is the relative discharge state of the battery counted from the fully charged condition (0.0 is fully charged), and t is time in hours.

do this really effectively would require connections to each cell with all the problems created by the many extra wires required. It has therefore been usual to divide the battery into blocks for monitoring purposes and to detect discrepancies by comparing the voltage and temperature of these blocks to others that are directly comparable. It is possible to detect gross cell faults by this method, but discrepancies in the SOC or condition of individual cells that can have a significant effect on the overall battery performance are not easily detected. Recent developments have made it possible to place voltage and temperature sensors on each cell and to transmit their data over a single wire digital data link to the battery management computer, making it possible to adjust the charging algorithm to ensure that undercharged cells are brought up to a fully charged state. EXtend Computer & Instrument Company of Walled Lake, Michigan, has extended this principle to include a bypass function in the battery sensor that allows the battery management computer to manipulate the SOC of the individual blocks or cells of the battery to keep the battery in optimum balance.

Detailed measurement and control of the condition of the cells of a battery has the potential to improve battery service life and performance, but the electric-vehicle designer has to balance the improvement obtainable with the cost of such a system.

7.7 Connection methods

When it becomes necessary to recharge the batteries of an electric vehicle two methods of connecting the battery within the vehicle to the outside power source are possible. One uses conventional conductive plug and socket coupling, the other uses inductive coupling through a specially designed transformer. In this transformer the primary and secondary windings can be taken apart so that the primary winding can be connected to the AC power supply either directly or through a frequency converter, and the secondary can be mounted on the vehicle and connected through the on-board charger to the vehicle battery. A generalised schematic diagram of such a system is shown in Figure 7.10.

The power to be transferred by the connectors is determined by the charging rate required. In the domestic situation the house wiring and the electricity supply company's infrastructure limits the power that can be conveniently supplied to a maximum of about 6 kW. In fact, if a large number of houses in an area were all charging their electric vehicles at the same time, for example overnight, the power it would be possible to supply without reinforcing the infrastructure would be limited to less than 3 kW. This is because the domestic power distribution system is designed on the basis that not everyone will be using large amounts of power at the same time, and in the UK it is usual to allow an average power level per house ('diversity factor'), of under 3 kW.

For the supply of power at these levels either conductive or inductive coupling can be used, it is when special fast-charging stations designed to recharge batteries in a few minutes are being used that the capability and design of the connectors becomes critical.

An idea of the connection options possible using conductive coupling is given in Figure 7.11. Taken from [8], this shows examples of the various power sources and infrastructures that could be used by the electric vehicle user in the course of a few days of general use. The diagram shows that without feedback of information on battery voltage, current, state of charge

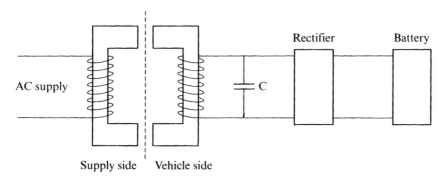

Figure 7.10   Basic layout of an inductively coupled power supply system

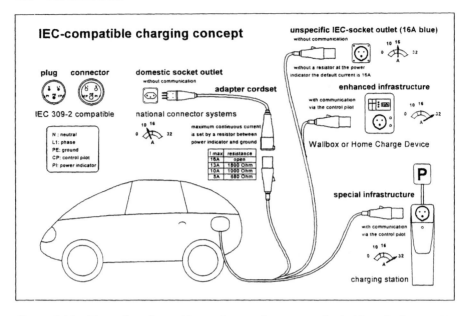

Figure 7.11   Examples of possible conductive charging methods (from Reference 8)

and temperature to the off-board charger via the control pilot link, it is necessary to limit the charging current to a safe 16 A. With communication it is safe to charge at up to 32 A since voltage and current can be fully controlled, particularly during the later stages of charging where close control of the charging voltage is required to avoid the battery overheating.

With specially designed connectors much higher currents can be used. This makes it possible to rapidly charge large battery systems on vehicles such as buses or to permit fast (15-min) charging of lead-acid or nickel-cadmium electric car batteries where this is necessary to enable long trips to be undertaken. Depending on battery capacity, voltage and the percentage of full charge that can be achieved after a 15-min fast-charge, currents as high as 200 A to 500 A may be required. Under these conditions feedback control of the charger, particularly in respect of battery temperature, is essential. It is also essential to keep the connector extremely clean and to maintain high-pressure contact between the male and female contacts, as at these high currents a high resistance contact can cause high contact temperatures and arcing can readily occur.

It has been suggested that it would be possible to design a charging system in which power is picked up from continuous embedded contacts in the highway. These contacts would extend over sufficient distance to form a 'charging zone'. As the electric car passed through this region an on-board contact arch would be lowered from the vehicle to make contact and charging would take place at a high rate during the limited time of contact possible. Accurate positioning of the vehicle would be essential as two sets of contacts would be required to complete the circuit. The idea seems to be vulnerable to a number of problems, not least those of contamination of the road surface by oil, water, or ice, causing poor contact and possible short-circuiting. Electromagnetic interference from the inevitably sparking contacts also seems likely to be a significant problem.

To avoid the problems of direct electrical contact and to minimise the risk of electric shock particularly in wet weather, inductive coupling has become a popular option. The general arrangement of an inductive power coupling is shown in Figure 7.10 with its realisation as a practical device currently in use in the General Motors EV1 electric car shown in Figure 7.12. The coupler unit is connected to the off-board charger and has a primary winding around a circular ferrite core in the form of a semi-circular 'puck' with a handle. This makes it easy to insert the primary winding into a receiving inlet slot and position it accurately against a corresponding secondary winding to receive the high-frequency AC power being transmitted by the off-board power supply. A block diagram of the complete charging system is shown in Figure 7.13. The system includes a radio link with antennae in the coupler and the inlet to transmit critical battery information on battery voltage, current, state of charge and temperature, back to the off-board charger to permit optimisation of the charging characteristic.

Inductive power coupling may be done at the public power supply frequency (50/60 Hz), [9] but at such a low frequency it is necessary to use a substantial magnetic circuit with a large coupling cross-section if efficient transfer of energy is to be obtained. It is also necessary to use a large number of turns on both the transmitting and receiving inductors. However,

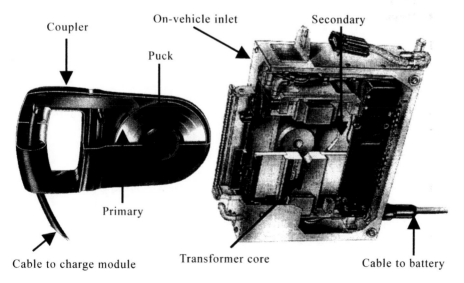

Figure 7.12   EV1 inductive coupler and vehicle inlet (from Reference 8)

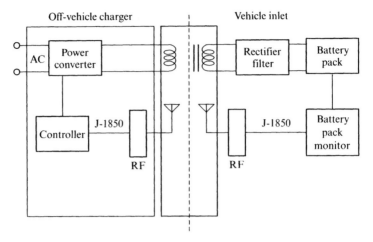

Figure 7.13   Block diagram of EV1 inductive coupling (from Reference 8)

if the public power supply frequency is converted electronically to provide power transmission in the 50 to 150 kHz region then it is possible to use relatively small inductors and highly efficient ferrite magnetic circuits as has been done in the EV1 inductive charger. With this high-frequency AC (HFAC) power with all its advantages, also come the potential problems of electromagnetic radiation at the power frequency and its harmonics. This is of particular concern in Europe where the low frequency band from 150 to 270 kHz is used for AM broadcast radio. It is therefore a high priority in the design of this type of system to protect against the problems that could arise from this radiation.

As in the case of conductive coupling there have been proposals to extend the inductive-coupling principle to enable vehicles to pick up power from inductors in the road surface. This is perhaps practical for a stationary vehicle but it is difficult to see how it could be made possible for a moving vehicle, if only for the requirement for a continuous inductor at or close to the road surface over long distances. The vehicle would also have to be driven carefully along a particular 'charging' lane. It would be necessary for the system to work at low frequency even with the vehicle inductor lowered as far as possible, since there would still need to be a large gap in the magnetic circuit to permit adequate vehicle ground clearance. An interesting side issue is the vulnerability of such a system to conductive material such as contaminated water or metal parts falling onto the road surface above the power inductors, or, in a traffic hold-up, vehicles with metal floor pans behaving like a short-circuited turn when stopped above the inductors.

There are also suggestions that it might be possible to recharge moving electric vehicles by means of radiated power from high-power roadside transmitters. It seems unlikely that such a system would be acceptable to the general public in view of their increasing sensitivity to even the very low radiated power levels from power lines and mobile phone systems.

It seems likely that both conductive and inductive connection methods will continue to be used for both normal and fast-charging of electric vehicle batteries. The method being determined by the relative importance individual electric vehicle manufacturers give to the issues of practicality, cost, safety, electromagnetic interference, efficiency of energy transfer and the customer-friendly aspects of the two methods.

7.8 Battery exchange

One method of overcoming the slow charging rate problem is to exchange batteries, and this requires the provision of special battery exchange stations at which discharged batteries can be removed from the vehicle and replaced by fully charged batteries held in stock. The discharged batteries are then recharged and used for further replacement.

The first requirement in such a system would be that the batteries would be hired from the exchange battery supplier with a guarantee of condition, rather than being owned by the vehicle operator, since if batteries were owned no one would wish to receive an old battery in place of their brand new one. The second requirement would be for the vehicle to be designed to make removal and replacement of a battery weighing up to 0.5 tonne easy and fast. This is difficult when safety legislation and crash testing legislation in particular is taken into account. Maintaining good high-current connections to the battery is also essential if dangerous overheating and arcing at the terminals is to be avoided.

Another financial rather than technical issue is the cost of carrying a large enough stock of batteries and the cost of the high-capacity charging system required. As an example, a large urban gasoline station would refuel on average about 1 500 vehicles a day. Transposing this to electric vehicles, and assuming each would require a replacement lead-acid battery worth, at a lowest estimated cost, about $2 000, implies an investment cost of $3 million for one day's stock of batteries (advanced battery types would be considerably more expensive). A minimum of two days' stock and possibly more is required to accommodate battery failures from age or misuse. Add to this the cost of a charging system to charge overnight this number of batteries together with the other required infrastructure and a total investment of at least $8 million is necessary. Although the batteries are recycled over a period of perhaps two to three years, obtaining an adequate return on this investment looks challenging.

The conclusion has to be that battery exchange is not likely to be a viable method of supplying power quickly for electric vehicles, although it may have a place under some specialist circumstances.

7.9 Infrastructure implications

If battery-driven electric vehicles are ever to be introduced on a large scale there will have to be three different forms of charging. The most widely used will be the domestic charger installation capable of a maximum charge rate of 6 kW, but normally used for overnight charging for eight to ten hours at a rate of 3 kW. Large numbers of electric vehicles charging overnight at this rate are no problem for the power station, and could provide a welcome load-levelling effect in a period of the day when other large loads are switched off. However, the local power supply wiring in an area where almost every house has an electric car charging overnight will be at its limit even at the 3 kW per car charging rate and may require reinforcement. As the period of charging progresses, the power required from the charging system will tend to decrease, but since the chargers are usually controlled by non-linear power-switching devices such as diodes and silicon-controlled rectifiers (SCRs), harmonic-current distortion is imposed on the power supply system. This can vary from 5 per cent to 100 per cent depending on the point in the charging algorithm that has been reached. Harmonic-current distortion can have a serious effect on the accuracy of power metering and even on the overall system stability [10]. The distortion is mitigated if a number of electric vehicles are charged at the same charging point since generally their batteries will be at different stages in their charging cycle due to their differing SOC. This causes the net current demand, power factor and magnitude and phase angle of the harmonic current distortion caused by each charger to be different leading to some cancellation. The authors of [10] have shown that the total harmonic distortion decreases as the number of electric vehicles being charged per station increases and converges to an average of about 25 per cent. An infinite number of electric vehicles at the same charging point would be required to reach this level; however, it can be closely approached if at least ten electric vehicles are charged at once. It is not clear how effective this process would be in the case of ten electric cars being charged in adjacent houses, or what the effect of the harmonic distortion would be on the domestic power supply system.

The second form of charging is known as 'opportunity charging' or 'boost charging'. This requires the availability of public charging stations (see Figure 7.14) wherever electric cars are parked, in particular at supermarkets, railway stations, in work place car parks and even along residential roads, long used as car parks by a large proportion of drivers. It seems unlikely that these charging stations could operate at high charging rates, 6kW being a probable maximum, since the cost of the large number of charging points required and the utility supply network to support them would be very substantial. Cost would be reduced if all electric cars were required to have on-board chargers, and this may be essential anyway if the differing charging requirements of the various battery technologies are to be accommodated.

(a)

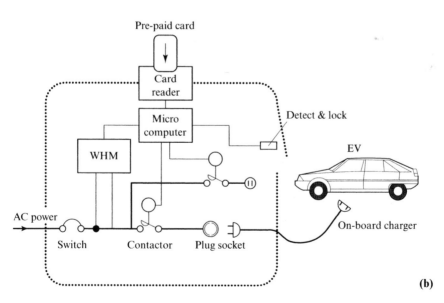

(b)

*Figure 7.14 (a) Public charging station (Kansai Electric Power Co.) and
(b) charging station circuit (Kansai Electric Power Co.)*

The third form of charging is the roadside fast-recharging station that serves the same function as the large roadside gasoline station does for conventional vehicles. Taking a charging time of 15 min to recharge a suitable battery from 20 per cent SOC to 80 per cent SOC as the fastest rate attainable with existing battery technology, and the average battery capacity of an electric car to be 20 kWh, neglecting efficiency losses, the energy transferred to each electric car battery to be charged will average 48 kW for 15 min. For comparison, when the tank of a gasoline-powered car is filled from empty, energy is transferred at the rate of more than 10 MW during the 90 s or so that this takes.

Assuming a charging station has the capacity to charge up to 20 cars at the same time (a not unreasonable number for a roadside facility) implies that provision would have to be made for a maximum utility power supply of more than 1 MW to allow for the higher charging rate in the early part of the charging cycle. This would be reduced to some extent by the likelihood that the vehicle batteries being charged would be on different parts of the charging cycle so that this demand would usually average out to a lower level. An alternative which would enable the power supply requirement to be reduced significantly would be the use of some method of energy storage capable of levelling load at least over a 15-min period. A possible candidate for this is a large stationary flywheel battery. Feasibility studies for this type of application are underway at the University of Texas in their High Power Energy Storage Program.

If electric cars are ever to be widely used, the provision of widespread battery-charging facilities is essential and substantial investment and further development of both chargers and the power-supply infrastructure will be required to meet the requirements outlined above.

7.10 Recharging/refuelling of other power storage devices

Power storage and/or generation in metal-air cells and in hydrogen/oxygen fuel-cells is fundamentally different from that in conventional chemical batteries. In the case of aluminium-air and zinc-air cells, under discharge the potassium hydroxide electrolyte absorbs oxygen from the air, which in turn causes erosion of the metal electrode and contamination of the electrolyte by oxides of the metal eroded. Recharging is normally done by replacing the metal electrode and the electrolyte, with the used electrode and electrolyte being reprocessed for future use. Because of the high-energy density of the metal-air cells (over 200 Wh/kg) and their suitability when combined in an electric vehicle battery for long-range operation (up to 400 miles on one charge has been claimed), interest in their use has been considerable and ingenious methods for replacing their electrodes have been developed. Electric Fuel Ltd in Israel have developed replaceable zinc cassettes which can be quickly exchanged, while Metallic Power of Carlsbad, California have developed a system using zinc pellets. Although so far only developed for powering small electric vehicles such as lawnmowers, the Metallic Power system appears to have potential for use in electric cars when fully developed. The system consists of a stack of zinc-air cells together with an electrolyte management system consisting of an air blower, electrolyte pump and control electronics. In operation the potassium hydroxide electrolyte is continually circulated between the fuel stack and the electrolyte management unit where the zinc oxide reaction product is removed from the electrolyte and stored. When refuelling, this zinc oxide is removed and zinc pellets are added to the cell stack. Effective recharging rates up to 1 MW are claimed to be feasible for a zinc-air electric car battery based on this principle.

In the case of hydrogen/oxygen fuel-cells, two alternative methods of refuelling are possible. One method is to refill the vehicle with pressurised hydrogen at a specially equipped filling station, the hydrogen then being stored in compressed or cryogenically cooled liquid form in a pressure tank, dispersed in metal hydride beds or possibly, in the longer term, absorbed in graphite nano-fibres. The second method is to generate the hydrogen on-board the vehicle using a 'reformer'. This reforms hydrocarbon- or alcohol-based fuels, usually methanol or gasoline, into hydrogen and water and is described in detail in Chapter 11. In either case refuelling is comparable in speed to that for a gasoline-powered car, although special safety precautions are necessary at the filling station when transferring hydrogen at high pressure because of the fire hazard. Some care also has to be taken when filling with methanol since it can be absorbed through the skin and has the potential to affect the nervous system.

As can be seen from the description of the recharging requirements for conventional chemical batteries in this chapter, there are no cheap and easy solutions that will overcome the problems of the slow recharge times and restricted operating range when they are used in electric vehicles. In the case of metal-air batteries and fuel-cells, refuelling rates and operating ranges comparable to those for gasoline-powered vehicles are possible and this is a major reason for the considerable interest in these methods of power storage at the present time.

References

1 HOBBS, R., KARMER, D., FLEMMING, F., and NEWNHAM, R.: 'Development of optimized fast-charge algorithms for lead-acid batteries', SAE Paper 1999-01-1157, 1999
2 US Patent 5949219, 'Optical state-of-charge monitor for batteries', US Dept. of Energy, 7 September 1999
3 US Patent 4460870, 'Quiescent voltage sampling battery state-of-charge meter', Curtis Instruments, 17 July 1984

4   US Patent 3562634, 'Method for determining the state-of-charge of nickel cadmium batteries by measuring the Farad capacitance thereof', US Atomic Energy Commission, 9 February 1971

5   US Patent 3984762, 'Method for determining battery state-of-charge by measuring AC electrical phase angle change', US Secretary of the Army, 5 October 1976

6   US Patent 4743855, 'Method of and apparatus for measuring the state of discharge of a battery', 10 May 1988

7   STEFFENS, W. and WESTBROOK, M.: 'A microprocessor based method of battery charge measurement', IMechE Paper C204/85, International Automotive Electronics Conference, Birmingham, October 1985, pp. 309–15

8   HAYES, J.: 'Battery charging systems for electric vehicles', IEE Colloquium – Electric Vehicles – A Technology Roadmap for the Future, London, May 1998

9   MILLS, J.: 'Inductively coupled battery charging system', MIRA Electric Vehicle Technology Seminar, April 1992

10   CHAN, M., CHAU, K., and CHAN, C.: 'Modeling of electric vehicle chargers', 24th Annual Conference of the IEEE Industrial Electronics Society 1998 (IECON '98), pp. 433–38

Chapter 8

Vehicle design and safety

So far in this book we have described the various subsystems which go to make up an electric vehicle but, with the exception of the hybrid electric vehicle, have not considered how all these systems come together to make a complete vehicle. This approach was necessary to enable the reader to understand the functions of the various subsystems sufficiently well to appreciate the limitations they impose on overall vehicle design.

Just as in a conventional vehicle, the size, weight, physical shape and interior space in the vehicle determine its appeal to the potential customer, as does the performance on the road and, particularly in the case of the electric vehicle, the time to recharge. As an offset to the long charging time problem, which was discussed in some detail in Chapter 7, electric vehicles often offer more flexibility in positioning of the subsystem components. This is because the propulsion electric motor or motors may be mounted close to or on the driven wheels with a flexible electrical connection to the battery and control system, instead of requiring the hard mechanical connection between engine and driven wheels necessary in a conventional IC-engined vehicle.

8.1 Effect of battery weight and volume

Any vehicle's performance is significantly affected by its weight and this is a particular problem in electric vehicles in which the only source of power is the battery. To obtain reliably what the user would probably consider to be a minimum acceptable range of 100 km (62 miles) with a typical small electric car would require over 400 kg of lead-acid batteries, about 200 kg of nickel-metal hydride (NiMH) batteries, or about 120 kg of lithium-ion (Li-Ion) batteries. This assumes that the battery is fully charged at the start and is discharged to the lowest practical level of 20 per cent state of charge (SOC) by the end of the journey.

In the electric car being considered, a lead-acid battery and its associated electric motor and controls is about twice the weight of the equivalent internal combustion engine, drivetrain and fuel in a conventional car. There is, however, a compounding effect of this additional weight which means that stronger and therefore heavier structural components must be used to support the concentrated battery weight and provide adequate crash protection. A rough rule of thumb is that for each additional kilogram of subsystem weight at least 0.3 kg of structural weight must be added. This results in an overall increase in the kerb weight of the vehicle of about 20 per cent and a corresponding loss of performance. This increase is reduced or eliminated when advanced batteries are used, but only if the same limited range, as is unavoidable with the lead-acid-powered vehicle, is accepted. If advantage is taken of the better energy density of the advanced batteries to use more batteries and increase range, the weight disadvantage of electric drive is not eliminated. In specially designed electric vehicles the use of lightweight materials, improved aerodynamics and sophisticated electronic controls can produce vehicles with comparable performance to their internal combustion-engined equivalents but cannot remove the severe range limitations caused by the low energy density of batteries compared to that of gasoline.

Another major problem in vehicle design is caused by battery volume. Compared to a lead-acid battery a tank of gasoline has more than 40 times the useful specific energy per kilogram and more than 20 times the useful energy density per litre of volume. This means that both the weight and volume of batteries must be correspondingly much larger than the fuel tank of a conventional car. In practice this has meant that many electric cars are only able to carry two people because of the space required for batteries. Advanced batteries improve this situation to some extent. Typically, NiMH batteries currently require 40 per cent less volume than lead-acid and Li-Ion over 60 per cent less for the same stored energy. In the case of Li-Ion a further advantage is the potential ability to form different-shaped batteries because of the flexible foil construction used.

8.2 Designing for minimum weight

For the reasons given above it is very important to design an electric vehicle to have the lowest possible weight. An example of this is Honda's 'Insight' hybrid in which extensive use is made of extruded, stamped and die-cast aluminium components and ABS composites to reduce body weight by 40 per cent below that of a comparable steel body.

In the case of a hybrid the weight of the internal combustion engine part of the system is also critical, and Honda claim to have achieved a 30 per cent reduction in weight by the use of slim case-hardened forged-steel connecting rods, a thin-sleeved block and extensive use of aluminium, magnesium and plastic for other engine components. The problem with the use of these advanced weight-reduction techniques is that they inevitably increase the vehicle cost significantly and make what is already an expensive vehicle even more costly. At the time of writing those manufacturers who are producing production hybrids have to subsidise their true cost by more than 50 per cent to bring the cost down to a level at which the general public will be prepared to lease or buy.

8.3 Safety of batteries

A major issue in electric vehicles is their crash performance. Not only does the concentrated weight of the battery present a mechanical problem under crash conditions, but there are also electrical problems that are unique.

The battery is normally placed in a specially designed compartment with sufficient strength and reinforcement to fully contain the battery under the front and side impact conditions specified in the various national regulations as shown in Table 8.1. Under these impact conditions the battery must not break loose or develop any electrical fault and, in particular, there must be no short-circuiting either internally or externally. In the case of some of the more exotic battery types, particularly the high-temperature (350 °C+) systems such as sodium-sulphur and sodium-nickel chloride, it is necessary to use special self-sealing battery cells to avoid the spilling of the hot and corrosive battery materials. Demonstrations of these high-temperature batteries still operating after a metal spike has been driven through them have been given and are testament to how effective this technology can be. However, it is probably better to use less unpleasant battery materials and operate at ambient temperature if possible.

In the case of some battery systems, particularly lead-acid, gassing under certain operating conditions is possible and has to be dealt with by providing adequate fan capacity for ventilation of the battery compartment. This is less of a problem with most advanced batteries but if high-rate charging is used a fan or even water cooling may be required. Generally, maintaining the battery at an optimum temperature helps considerably in obtaining long

Table 8.1   Car impact test requirements

	Frontal barrier impact (km/h)	Offset frontal impact (km/h)	Side impact (km/h)
Japan	50		50
Europe	64 (1)	56 (1) (40% offset)	48 (1)
USA	56 (2)	64 (3)	54 (combined) (2) (48 km/h bullet veh. 8 km/h target veh.)

(1) European New Car Assessment Programme (EuroNCAP)
(2) US National Highway Traffic Safety Administration (NHTSA)
(3) US Insurance Institute for Highway Safety (IIHS)

life and is particularly important in the case of high-temperature batteries such as sodium-sulphur and sodium-nickel chloride. In the high-temperature case, sealed battery containers with facilities for electrical heating and for evacuating the air within the container are necessary. In the case of batteries intended to operate at ambient temperature heating may have to be provided if the vehicle is to be operated regularly in freezing temperatures.

The battery voltage also presents a significant safety problem. Any voltage above 50 V from a low impedance source can give a potentially fatal electric shock if the resistance of the connections to the person concerned is sufficiently low to permit a high enough current to flow through the vital organs. However, in an AC system 50 V peak level corresponds to only 35 V RMS, making it even more difficult in an AC system to keep to an intrinsically safe voltage level. In practice, an electric vehicle operating with a 50 V system would require a large and heavy electric motor and heavy wiring, switching and contacts to accommodate the very high currents which would be required. To avoid this and to permit the use of much lighter-weight components, it is normal to use voltages up to 350 V in cars and up to 500 V in commercial electric vehicles. When these higher voltages are used, great care has to be taken to insulate all cables and connectors from any possibility of contact by operators or service personnel and to protect these cables and connectors from any contact in a crash. It is usual to isolate the electrical system from the vehicle chassis so that any short to chassis anywhere in the system can be detected and the system shut down by protective devices and protected as far as is possible under an emergency procedure. Interlocks are also provided so that high-voltage circuits are disconnected whenever vehicle compartments containing these circuits are opened. Isolating the electrical system from the chassis also reduces the danger to operators or service personnel if accidental contact should take place, since that point in the system where the contact is made becomes the only point on the system at the same potential as the person or object making the connection. Nevertheless, it is essential to treat systems at full battery voltage with great respect. It should also be remembered that a large amount of energy is stored in a charged electric vehicle battery (as it is in a tank of gasoline), and any short-circuit which does occur in an accident could result in a major release of this energy and a consequent explosion and fire.

Since batteries will consist of many low voltage cells connected in long series strings, and some of these strings may also be paralleled, there are many connectors and interconnecting cables required. The resistance of the connectors and cables combined with the internal resistance of each cell determine how much variation there is in the voltage at each cell during charge and discharge and therefore whether the cell will be under- or over-charged during operation. It may be necessary to monitor individual cell temperatures and voltages, as discussed in detail in Chapter 7, if maximum power and life are to be obtained from a battery system. The inductance of the cable system may also be a factor in the total circuit inductance and must be allowed for in the design of the associated electronic control circuits.

8.4 Safety of alternative energy generating and storage systems

With energy generating and storage methods such as fuel-cells, flywheels or hydraulic accumulators different safety issues arise. In the case of fuel-cells the safety issue is largely that of containing the hydrogen fuel safely. The simplest system is to store the hydrogen in a liquid form in a pressure tank. The tank needs to be about four times the size of a gasoline tank in a conventional vehicle if the same range on one filling is to be achieved and it is vulnerable to rupture in a severe crash with consequent risk of fire and/or explosion. An alternative is to cool the hydrogen to a liquid form in a cryogenically cooled tank, although this requires additional energy expenditure. A second and better alternative is to disperse the hydrogen in metal hydride beds as this limits the rate at which hydrogen can be extracted and therefore reduces the risk of fire and explosion in an accident. The recharging of any of these on-vehicle hydrogen storage systems involves the difficult task of providing a safe method of delivering suitably processed hydrogen at filling stations in such a way that users can refill their tanks easily and safely. Most of these safety problems can be overcome if, instead of storing substantial quantities of hydrogen on board, it is generated by a process known as 're-forming' as required.

Reforming involves the steam reforming and/or partial oxidisation of a suitable liquid hydrocarbon or alcohol fuel, which can be stored in a conventional gasoline tank. Physically compact reformers are now available (see Chapter 11) which together with the fuel-cell system do not take up more space than a conventional electric vehicle battery. Current work [1] is aimed at packaging the whole fuel-cell/reformer system under the floor of a car so that the passenger and luggage space is not compromised in the way that it currently is in most EV installations. The quantity of hydrogen that is generated by the reformer is relatively small at any one time during operation so that the risk of fire or explosion is small. Currently, most work has been done on systems of this type using methanol as the hydrocarbon fuel although there is work in progress to use gasoline. This would make it possible to use the existing fuel infrastructure without any modification or additional safety implications and avoid the care required in handling methanol because of the risk of it being absorbed through the skin where, in sufficient quantity, it can affect the nervous system. Methanol can also cause severe corrosion where it is in contact with vehicle components and it has been suggested that there could be a risk of pollution of water supplies if it was in widespread use at filling stations.

In the case of the flywheel the main safety concern is the risk of serious physical damage if the flywheel is not properly contained in an accident.

Large amounts of energy can be stored in a high-speed flywheel and if this is released suddenly it can be extremely dangerous. The use of composite materials such as carbon-fibre for high-speed flywheels instead of conventional solid metal, reduces the danger considerably; if a carbon-fibre flywheel bursts it will normally disintegrate into small fragments which behave like a fluid inside the containment vessel, equally distributing the energy to be dissipated. This is in contrast to the solid-metal flywheel that will usually break into large pieces that then develop very large stresses over small areas of the containment vessel. It is therefore extremely difficult to design containment for solid-metal flywheels which is not impracticably large and heavy.

Physical damage and release of energy is also the main safety issue with the hydraulic accumulator in an accident situation. However, it is not possible to store a large amount of energy, so that the consequences of failure are potentially less severe than in the case of the flywheel.

8.5 Battery disposal and recycling

Of the materials used in battery construction, some are safe to handle and others have properties that make them dangerous to handle and complex to recycle. The materials used in lead-acid batteries have been recycled for many years and it would be straightforward to extend the existing recycling system to deal with lead-acid electric vehicle batteries. On the other hand, the cadmium in nickel-cadmium batteries is particularly toxic, as is the lithium in lithium-iron sulphide, lithium-solid polymer and lithium-ion batteries. The sodium in sodium-sulphur and sodium-nickel chloride batteries is also, if not toxic, still very unpleasant to deal with, particularly if it is at the operating battery temperature. It is undoubtedly true that the success of most of these advanced batteries has been seriously inhibited by these problems. For example, nickel-cadmium batteries have been used quite extensively in experimental electric vehicles particularly because of their good energy density and capability of accepting high rates of charge. However, they are now being replaced generally by the higher energy density nickel-metal hydride battery which uses relatively benign materials and can also accept high charging rates.

8.6 Safety of other electrical systems

Electrical systems other than the battery and its connectors also require protection from various problems that may arise during operation. In particular, protection of power transistors against overvoltage and short-circuiting is normally provided within the electronic controller by the use of 'snubbers' (see Chapter 4) and fast fusing or circuit breakers. Thermal protection is also necessary for the battery, motor and controller and this is provided by the use of temperature sensors in close contact with all the devices that may be affected. These sensors are then connected to the electronic controller so that the system operating conditions can be modified and forced cooling provided to reduce the temperature of the overheating component or system.

In cases where DC commutator motors are used, it is necessary to provide overspeed protection using speed sensors on the motor capable of opening the armature circuit through the electronic controller. This is not required in the case of induction or synchronous motors as this control takes place automatically through the frequency control required in normal operation to determine motor speed.

The auxiliary power system, which includes all the services required in a conventional car, such as lighting, instrumentation, heating, air conditioning etc., also requires protection and this is provided by conventional fusing and circuit breakers.

8.7 General design and safety issues

There are a number of design and safety issues that are peculiar to electric vehicles both because of the necessity of conserving battery energy for propelling the vehicle and because of its special characteristics. Examples of this are the provision of heating and air-conditioning subsystems, the maintenance of the auxiliary power subsystem mentioned above and the special requirements of the braking, suspension and wheel systems.

8.7.1 Heating and air-conditioning

It is normal practice in electric vehicles to provide heating using a heater burning a hydrocarbon fuel such as propane, gasoline or kerosene. Although this gets away from the ideal of a vehicle with no hydrocarbon or carbon dioxide emissions, these heaters can be made to operate very efficiently with minimum emissions. The alternative of using what waste heat there is available from the vehicle control system and drive motor and combining this with an auxiliary electric heater may be acceptable in mild climates. In the GM EV1 a 'heat pump' is used together with waste heat and electric heating but it is not known how effective this is in cold-climate use. It is important to minimise the energy taken from the battery when ambient temperatures are low as most batteries operate less efficiently under these conditions. Use of an electric heater for long periods will certainly reduce operating range considerably.

In the case of air-conditioning there is no easy solution. The only good thing is that in high ambient temperature conditions where air-conditioning is required, batteries are generally operating at a higher efficiency. The basic requirement in an air-conditioning system is to be able to adequately power the compressor pump and fan to reduce the air temperature inside the vehicle to an acceptable level within a reasonable time and then hold it at that level. The vehicle can be designed with better insulation and heat reflecting glass to reduce the radiant heat input and solar-cell-powered fans can be used to keep the inside temperature down when the vehicle is parked. Even when these strategies are used, however, it is unlikely that it will be possible to provide the level of air-conditioning that users have come to expect in conventional vehicles. This level of comfort can only be provided by using energy at a rate between 30 per cent and 50 per cent of that required to propel an electric vehicle under normal operating conditions. This implies a reduction of range of the same percentage if full air-conditioning is operating.

8.7.2 Auxiliary power subsystem

The auxiliary power subsystem can include a wide range of electrical features. In a conventional vehicle the user would expect to be provided with not only the heating and air-conditioning already mentioned, but also power steering and front and rear window defrost, as well as the essential lighting, instrumentation, windshield wipers, electric windows and door locks and other features. Some of these features require significant amounts of electrical power and to permit the use of mass-produced components must operate at 12 V. In the case of the electric car this requires the provision of a 12 V battery in addition to the high-voltage propulsion battery. This battery is normally kept nearly fully charged by a DC/DC converter from the main propulsion battery with auxiliary charging from solar cells being used in some cases.

If all the features listed above are included in an electric car, high currents will be required under some operating conditions. To minimise this, time controls with a severely restricted operating time are essential on some high-current features such as front and rear window defrost. However, lighting must be used at night and windscreen wipers in the rain so that significant use of power and therefore consequent reduction of operating range is unavoidable. High-efficiency lighting systems and improved small motor design can reduce these unavoidable loads to some degree, but the electric vehicle designer must always be aware of their significant effect.

8.7.3 Braking, suspension and wheel systems

In an electric vehicle it is usual to obtain braking by using the fact that the propulsion motor can also be used as a generator charging the battery and therefore can provide a strong retarding force on the driven wheels. The power generated is returned to the battery via the controller and can provide sufficient additional energy to increase vehicle range by between 5 and 15 per cent depending on the driving cycle being operated. It is also necessary to provide conventional friction braking to take care of the situation where the motor/generator is running at low speed and is unable to generate sufficient energy to brake the vehicle effectively. It might be thought that this additional regenerative braking ability would allow lighter, lower-cost mechanical brakes to be used. Unfortunately, this is not the case as these brakes must not only be able to stop the vehicle effectively if the electric propulsion system fails but must also cope with a vehicle which is usually of greater weight than its conventional equivalent. On the plus side, the life of the friction brakes is considerably improved since they need to be used less when regenerative braking is in use.

It is straightforward to provide regenerative braking when induction, synchronous or switched-reluctance propulsion motors are used since the motor will operate as a generator if the drive frequency is reduced. When this regeneration is taking place the induction motor rotor speed will be greater than the synchronous speed, while in the synchronous and switched-reluctance motors the rotor will be displaced ahead of its motoring position. The controller can operate bilaterally without any modification. It is more complex with a low-cost DC motor control system such as a chopper since it is necessary to reverse either the motor field or armature current so that it will operate as a generator. It may also be necessary to provide additional components in the controller to permit current flow to as well as from the battery. It is advantageous if a gear-changing transmission is used, as this will enable the drive train to operate at best efficiency in both drive and regenerative conditions.

The design of the suspension of an electric vehicle is very dependent on the weight and position of the main propulsion battery. A lead-acid battery may be up to one-third of the total vehicle weight with a consequent need for significant strengthening of the suspension components and balancing of their operation between front and rear to allow for the positioning of the battery. Since batteries are ideally positioned under the vehicle floor at as low a level as possible, the centre of gravity is lower than in a conventional vehicle, and this can improve handling when cornering. Advanced batteries such as NiMH and Li-Ion have lower weight and volume for the same specific energy and, unless they are used in larger assemblies to obtain greater range, will not require the same suspension reinforcement as lead-acid batteries.

Wheel systems are generally conventional but must accommodate high-pressure tyres. In future they may need to be designed to accommodate in-wheel drive motors.

8.7.4 Rolling resistance

To obtain maximum range with any vehicle it is necessary to not only make the vehicle as light as possible but also to reduce the rolling resistance of the vehicle to the lowest possible level. This is particularly critical with an electric vehicle because of the limited capability to store energy.

The rolling resistance (R) of a vehicle is expressed by the following formula:

$$R = (W \times \mu_r) + (S \times \mu_a \times V^2) + (W \times \alpha) \qquad [2]$$

Where

W = vehicle weight
μ_r = rolling resistance coefficient
S = frontal area
μ_a = drag coefficient (C_d)
V = driving velocity
α = acceleration rate

Vehicle weight, rolling resistance and drag coefficient are the factors in this formula that can be improved by design. In particular, reduction in the weight of the body, chassis and battery (and of the heat engine in the case of hybrids), reduction of the aerodynamic drag of the body and reduction of the rolling resistance of the tyres, can all contribute substantially to reducing the rolling resistance of the vehicle.

In Section 8.2 we described how Honda have used advanced materials and design techniques to reduce the weight of the body, chassis and engine of their Insight hybrid. The GM EV1 is also a good example of how an electric vehicle can be designed to have minimum weight, rolling resistance and drag coefficient. Body and chassis weight have been minimised by the extensive use of aluminium and composites, rolling resistance reduced by the use of high-pressure tyres running at more than twice the normal tyre pressure level and drag (C_d) reduced by careful body design to 0.19 compared to 0.24 for the best conventional cars. These reductions have reduced the energy consumption to 70 Wh/km at a steady 88 km/h (55 mph), about half that for a conventional small internal combustion-engined car. More recent electric vehicle designs have been claimed to have even lower energy consumption. Of course, most of the design techniques used to obtain these improvements could be used in conventional cars and development is taking place to use some of this technology to meet the ever more stringent average fuel consumption levels now being mandated in some countries. However, the incentive to spend the significant amount of money required to fully implement these improvements is not nearly so great for conventional vehicles as it is for electric vehicles. This is primarily because the range obtainable from each refuelling is not the major issue it is in the electric vehicle, but also because the highly desirable improvements in gradeability, acceleration and top speed which are also obtained can easily be provided in the conventional vehicle with the surplus power available from the internal combustion engine.

References

1   TRAN, D.: 'DaimlerChrysler fuel cell report' – Panel Presentation at NAEVI 1999 Atlanta, Georgia, November 1999, EV World – Conferences Report, http://www.evworld.com
2   'Electric vehicles – technology, performance and potential', International Energy Agency, OECD 1993

Chapter 9

Battery electric cars

In this chapter I have brought together all the information available on the current (2001) production, prototype and experimental battery electric cars developed by the major manufacturers together with more detailed descriptions of four of the production vehicles which I consider to be of major significance. This information is given in Tables 9.1 and 9.2. The technical details given have been provided directly by the manufacturers except in a few cases where information has been obtained from published papers and press reports. Specialist niche manufacturers of electric vehicles such as Bombardier and Solectria are not included, as it is my view that it is essential for the major automotive manufacturers to commit to electric vehicles for them to succeed. It is therefore the developments introduced by the major manufacturers that we should concentrate on if we are to understand where electric car development is heading.

9.1 Production electric cars

In Table 9.1 I have listed 11 electric cars which are currently or have until recently been on sale or for lease to the general public from the major motor manufacturers. The list is undoubtedly incomplete in spite of many enquiries within the automotive industry and also suffers from lack of data on some vehicles, but it does give some idea of the range of battery and drive technologies in use and of the top speed and all-important range of these production vehicles. It is interesting to note that of the battery types used, nickel-cadmium (four vehicles) and nickel-metal hydride (three vehicles) are the most popular choice. Lithium-ion is used in two vehicles and VRLA lead-acid in one. The battery type used is not specified in one of the vehicles listed. Functionality rather than cost has undoubtedly been a major factor in deciding which battery type to use for what is essentially small pilot-production intended to show electric vehicle capability. NiCd and NiMH advanced batteries are projected to be at least four times as expensive as

Table 9.1   Production electric cars

Manufacturer	Citroen	Daihatsu	Ford	GM	GM	Honda	Nissan	Nissan	Peugeot	Renault	Toyota
Model name	AX/Saxo Electrique	Hijet EV	Th!nk City	EV1*	EV1*	EV Plus*	Hypermini	Altra EV	106 Electric	Clio Electric	RAV 4
Drive type	Separately excited DC	PM Synch	3-phase induction	3-phase induction	3-phase induction	PM Synch	PM Synch	PM Synch	Separately excited DC	AC induction	PM Synch
Battery type	NiCd		NiCd	Lead-acid (VRLA)	NiMH	NiMH	Li-ion	Li-ion	NiCd	NiCd	Ni MH
Max power O/P (kW)	20		27	102	102	49	24	62	20	22	50
Voltage (V)	120		114	312	343	288		345	120	114	288
Battery energy capacity (kWh)	12		11.5	16.2	26.4		15	32	12	11.4	27
Charging connector			Conductive	Inductive	Inductive	Conductive	Inductive	Inductive		Conductive	Conductive
Top speed (km/h)	91	100	90	95	129	129	100	120	90	95	125
Claimed max range (km)	80	100	85	95	130	190	115	190	150	80	200
Charge time (h)	7	7	5-8	6	6	6-8	4	5	7-8		10
Sale or lease price	$12 300 excluding batteries	$23 990		$399 per month	$480 per month	$455 per month	$36 000 or $23 350 with subsidy	$50 999 or $599/month	$27 000 or $14 700 excluding battery	$27 400 or $16 000 with subsidy	$45 000 or $499/month

* Production suspended at time of writing (February 2001)

basic lead-acid and twice as expensive as Li-Ion, even in large production quantities (see Chapter 5, Table 5.1). NiMH batteries are also reported to be up to ten times as expensive in small experimental quantities. It is interesting to note that no high-temperature batteries are so far in use in production cars in spite of their long period of development and pilot production.

Drive systems, where these are known, are AC with the exception of the Citroen Saxo and Peugeot 106, which both use separately excited DC motors. AC drive systems are almost equally split between induction motors and permanent magnet synchronous motors. Battery voltage varies from a low of 120 V to a high of 343 V and maximum range from a claimed, very optimistic high of 200 km to a low of 80 km.

These statistics indicate how different the aims are of the manufacturers, a sure sign of the immaturity of the technology and the relatively small number of vehicles so far sold to the public. It is the author's opinion that it will take another ten years for any form of optimised design to evolve.

It is useful to look in some detail at some of these production electric vehicles as they are setting the agenda for the improved versions we shall see over the next few years. In particular the GM EV1, the Ford Th!nk, the Nissan Hypermini and the Toyota RAV 4 are representative of current technology and show both the potential and the limitations of the 'pure' electric vehicle.

9.1.1 The General Motors EV1

The EV1 was first seen at the Los Angeles Auto Show in January 1990 and was marketed in the USA from the autumn of 1996. The car, shown in Figure 9.1, is a two-seater specifically designed for commuting and urban travel, a usage pattern very appropriate to its limited range between charges. The specification includes dual air bags, anti-lock brakes, a CD player and cruise control. Because it is so quiet on the road it is also fitted with special chimes, light signals and a reversing horn to warn pedestrians of its presence.

When originally marketed EV1 was offered with a valve-regulated lead-acid (VRLA) battery weighing 530 kg and able to store 16.2 kWh of energy. This battery can deliver more than 450 deep discharge cycles and is claimed to give the vehicle a maximum range of 145 km. In 1998 GM introduced a nickel-metal hydride (NiMH) battery as a further option. This advanced battery can store nearly twice the amount of energy of a VRLA battery of similar size and weight and during development testing an experimental EV1 is reported to have covered 600 km on a single charge. NiMH batteries, although at least four times more expensive than VRLA also have up to four times longer life. Both types of battery have a fast-charge capability and can accept an 80 per cent recharge in about 20 min although normally recharge times would be in the 3 to 6 h range depending on the supply available.

Table 9.2 Prototype and experimental electric cars

Manufacturer	BMW	BMW	Daihatsu	Daimler Chrysler	Daimler Chrysler	Fiat	Ford	Ford	GM	GM	Lada	Mazda	Mazda	Mitsubishi	Peugeot	Toyota
Model name	E1	BMW Electric	Charade Social EV	Zytek Smart EV	A-Class Electric	Seicento Elettra	Think Neighbor	e-Ka	Impulse 3	Impulse 3	Rapan	Roadster-EV	Demio EV	Libero	Ion	E-com
Drive type	PM Synch	PM Synch		Brush-less DC	3-phase induction	3-phase induction	DC	3-phase induction	2×3-phase induction	2×3-phase induction	Separately excited DC	AC	PM Synch		Separately excited DC	PM Synch
Battery type	$NaNiCl_2$	$NaNiCl_2$	NiMH	$NaNiCl_2$	$NaNiCl_2$	Lead-acid		Li-Ion	NiCd	$NaNiCl_2$	NiCd	NiCd	NiMH + super-cap.	NiMH + NiCd	NiCd	NiMH
Max power O/P (kW)	32	45	30	30	50	30	5	65	45	42		30			20	19
Voltage (V)					289	216	72		210	286		192				288
Battery energy capacity (kWh)		29		13	30	13		28	15	26						
Charging connector						Conductive			Inductive	Inductive						
Top speed (km/h)	130	130		97	130	100	40	130	120	120	90	130		130		100
Claimed range (km)	140	155	120	160	200	90	48	150	80	150	100	180		250	150	100
Charge time (h)		8 (75% fast charge 40 min)			7	8 (80% in 4 h)	4–8	6	6	6–8	8	8				
Date for production			2000				Nov 2000								2000	

The propulsion system is a three-phase AC induction motor with integrated gear reduction capable of providing up to a peak of 102 kW to the wheels. The motor can operate as a generator on the overrun or during braking to provide regenerative recharging of the battery. This is claimed, perhaps optimistically, to extend the vehicle range by up to 20 per cent. The battery and motor system is controlled by an electronic module that also controls battery charging. This module also controls automatic and manual disconnection of the 312 V main battery supply for safety, as well as using battery temperature and voltage sensors to control battery thermal management and venting. The waste heat from the module (about 600 W) is used, together with the heat from a computer-controlled 'heat pump' system and an auxiliary electric heater to provide interior heating.

Charging using the on-board charger from the domestic electricity supply at 110 V takes up to 15 h, but if connected to a purpose designed 220 V 30 A charger as shown in Figure 9.2 three hours is sufficient for a full charge. Charging is done using a weatherproof inductive 'paddle' rather than through a conventional plug and socket and it is claimed that, if required, power can be transferred at a rate of up to 120 kW by this method.

As the first production electric car produced by a major automotive manufacturer with an on-road performance comparable to conventional cars, the EV1 has not only been a showcase for the latest electric vehicle technology but has also shown that electric vehicles can be produced by normal production methods. It is still too expensive at $33 995 to make it attractive to members of the general public who might consider an electric car as a suitable second vehicle for commuting and urban use, but it has shown what can be achieved at the present state of the technology. Unfortunately, production of the EV1 was suspended in April 2000 and those that were on a lease agreement were withdrawn. It was reported this was because of an

Figure 9.1   General Motors EV1 electric car

Figure 9.2   EV1 charger

unspecified problem with the charging system. It is not clear if production of the vehicle will be resumed.

9.1.2 The Ford Th!nk City

The Ford Th!nk City shown in Figure 9.3 is the latest production version of the small two-seater urban compact electric car originally developed by the Norwegian company Pivco Industries. This company was taken over by Ford Motor Company in 1998. The Th!nk City car has been in production since November 1999. As of December 1999, 120 cars had been ordered. A USA version of the Th!nk City is currently under development and was reported in January 2000 as being available within two years.

The car is a purpose-built two-seater, powered by a liquid-cooled three-phase AC induction motor with a maximum power output of 27 kW. This is sufficient to give a 0 to 50 km/h time of 7 s and a top speed of 90 km/h. Power is from a 114 V nickel-cadmium (NiCd) battery which is water-cooled for maximum performance and has a power storage capacity of 11.5 kWh.

The car has a range of 85 km when run on the EC test cycle, although of course as with all electric cars, real-world range is significantly affected by how the car is driven, by the terrain and by heavy use of the auxiliary electrical systems.

The Th!nk City uses a conductive plug and socket connector just under the windscreen to connect the on-board charging system to an external power supply and this connector can be plugged in to any standard 220 V 16 A domestic supply outlet. With this supply it takes about 8 h to fully recharge a discharged battery and 5 h to reach 80 per cent charge.

Safety has been of high importance in developing the design of the car. The frame consists of a high-strength steel undercarriage married to an aluminium space frame that forms the upper skeleton of the passenger compartment. This steel undercarriage is designed to absorb energy during a frontal collision. In side impacts, the 1.8 m wheelbase of the car means that most vehicles side-impacting this car will have their impact energy absorbed by the aluminium door beam, at least one tyre and the sealed compartment beneath the driver's seat where the 250 kg battery is stored. Driver's air bags and safety belt pretensioners are also provided and have enabled the car to perform as well in international crash-test programmes as conventional gasoline-fuelled cars.

The price at which this car is being offered to the public is not known, but if that price can be made competitive with similar small urban conventional cars it seems probable that a significant number of these cars can be sold. They are likely to be of wider interest than just for use in cities where zero-emission regulations apply.

Figure 9.3   Ford Th!nk City electric car

Figure 9.4   Nissan Hypermini electric car

9.1.3 The Nissan Hypermini

The Nissan Hypermini shown in Figure 9.4 is similar in many ways to the Th!nk City, being also a two-seater compact urban commuter car. It was due for domestic release in Japan in February 2000 and this release was preceded by the release of 35 cars for experimental use in park-and-ride and rental schemes.

The car is powered by a neodymium permanent magnet synchronous motor developing a maximum power of 24 kW, a maximum torque of 130 Nm and a top speed of 6 700 rpm. This motor, which drives the rear wheels, generates a broad torque curve from low speed up to the maximum vehicle speed of 100 km/h and is claimed to provide smooth and rapid acceleration with low noise and vibration. This smooth motor control is achieved by the use of a high-accuracy rotor-angle-detection sensor feeding motor speed information to a high-performance 32-bit microcomputer which is used to obtain optimum motor control matched to the driving conditions.

The motor is powered by a lithium-ion battery with an energy density of 90 Wh/kg and power storage capacity of 15 kWh. Because this battery is nearly three times more efficient at storing energy than lead-acid batteries and about twice as efficient as those of nickel-cadmium, it is possible to reduce the battery weight while still providing the same driving range. Li-Ion batteries also do not suffer from the memory effects present in nickel-cadmium batteries. The Hypermini has a range of 115 km when tested on Japan's 10–15 test mode using this high-performance battery and high-efficiency motor. However, range will be significantly reduced if the air-conditioning system which is fitted as standard is used for long periods.

Battery charging is by an inductive charging system similar to the GM EV1 and power is transferred by a non-contact electromagnetic induction 'paddle' inserted in a socket mounted in front of the windscreen. The 'paddle' is fed from a small 200 V charger which can be installed in a garage and connected to the domestic power supply. Full charge can be completed in approximately four hours.

The design of the Hypermini is based on a lightweight but rigid extruded tubular aluminium space frame. This is combined with aluminium panels and aluminium castings to produce a high-energy absorption body which is claimed to be over 20 per cent lighter, 40 per cent more rigid and having 20 per cent fewer parts than a comparable monocoque steel body. This body structure is excellent at absorbing crash energy, in part due to the four times greater absorption efficiency of aluminium compared to steel, but also to the use of straight structural members wherever possible and the adoption of large cross-section hexagonal front-side members.

At a price in Japan of 4 million Yen ($36 000), the Nissan Hypermini is unlikely to sell in large numbers, but hopefully it will be possible to produce future versions at prices that are more competitive with conventional small cars. The introduction of a zero-emission requirement in cities would, of course, make a vehicle of this type much more attractive.

9.1.4 The Toyota RAV 4 EV

Of the four production electric cars selected for detailed description, the Toyota RAV 4 EV urban electric vehicle shown in Figure 9.5 is perhaps the nearest to a conventional car in appearance and accommodation. Derived directly from the conventional RAV4 L, it has been possible to accommodate the batteries under the floor so that, in contrast to the first three cars described, there is a roomy cabin and a large luggage space. The RAV 4 EV was put on sale in the Japanese automotive market in September 1996.

Propulsion is by a compact lightweight permanent magnet synchronous motor driving a transaxle through a two-stage reduction gear unit. This motor has a maximum power output of 50 kW over the speed range 3 100 to 4 600 rpm, develops maximum torque of 190 Nm between 0 and 1 500 rpm and propels the vehicle at a top speed of 125 km/h.

The batteries are high-performance sealed nickel-metal hydride (NiMH) weighing 450 kg and consisting of 24 sealed 12 V 95 Ah units capable of storing up to 27 kWh of power. Maximum range on a single charge is claimed to be 200 km on the USA combined test cycle.

Since this is not a specially designed electric car, but rather an adaptation of an existing vehicle, no information is available on what changes, if any, have been made to the structure of the vehicle to accommodate the electrical drive system and batteries.

Figure 9.5   Toyota RAV4 electric car

The price in Japan of the RAV 4 EV is understood to be $45 000, too high a price for anything but a niche market. If electric cars are ever to become significant sellers in the automotive market, prices will need to be at about half their existing level, and this seems likely to involve a greater leap of faith in the technology than major manufacturers are prepared to take in production cars at present.

9.2 Prototype and experimental electric cars

Table 9.2 lists 16 prototype and experimental electric cars that interestingly use a range of battery technologies. Lead-acid batteries only appear once in spite of their relatively low cost, but NiCd is used in five vehicles, NiMH in three and Li-Ion in one, this being the recently announced Ford e-Ka. Five vehicles (three manufacturers: GM, BMW and DaimlerChrysler) use the high-temperature technology of sodium-nickel chloride ($NaNiCl_2$). It is difficult to detect any clear preference among the manufacturers represented and this emphasises the experimental stage most are still going through.

Chapter 10

Hybrid electric cars

Any vehicle that has more than one power source can be classified as a hybrid electric vehicle (HEV), but most frequently the term is used for a vehicle which combines electric drive with a heat engine using a fossil-fuel energy source.

The heat engine-electric hybrid vehicle, first conceived as long ago as 1900, is intended to overcome the limited range and long recharge time of the 'pure' electric vehicle which relies for its energy source only on its battery. It usually achieves this desirable objective by combining electric drive with a conventional heat engine using a fossil fuel energy source. This engine may use any of a wide range of heat-engine technologies including two- and four-stroke spark-ignited gasoline engines; two- and four-stroke diesel engines; Stirling engines; or gas turbines. These engines are designed for operation at maximum efficiency to reduce fossil-fuel consumption and emissions to the lowest possible level, and this is achieved by the use of various methods of combining the two propulsion systems and by sophisticated control strategies to optimise their interaction.

The basic conventional automotive drive system uses an internal combustion engine coupled to the transmission and the road wheels through a gearbox that is used as a matching device between the engine and its load. To provide sufficient power at the road wheels for adequate acceleration and hill climbing (gradeability) at an acceptable speed, it is necessary to use an engine with a maximum power output about ten times greater than would be required to propel the vehicle at the same speed along a level road. This makes it impossible to operate the engine efficiently at the relatively low power required under most normal operating conditions.

Significant improvements to the efficiency of the conventional engine and transmission system can be made by the use of an electronically controlled continuously variable transmission (CVT), variable valve-timing and a tuned variable-inlet manifold. When these features are combined with optimised electronic control of fuelling and ignition timing, it is possible to ensure that the engine operates much closer to the optimum conditions over a wide range of speed and load conditions. Unfortunately, however, these improvements in efficiency are limited by the relatively low efficiency of the internal combustion engine and those continuously variable transmissions developed up to the present time. It is also not possible to recover energy lost during braking, in contrast to battery electric and hybrid electric vehicles in which energy may be returned to the battery during braking or coasting. In the case of the hybrid electric vehicle it is also possible, depending on the way in which the two power sources are coupled and controlled, to operate the heat engine at close to constant speed and load and therefore under high efficiency, low emission conditions.

Hybrid electric vehicles are ideally operated in a mode in which they run under battery power where this is sufficient to meet the requirements of the terrain and driving style, with the heat engine being automatically started and run when the power required cannot be met by the electric drive, or the battery has discharged sufficiently to require recharging. From the point of view of vehicle weight it is desirable that the battery is as small as possible to reduce the energy used simply to transport the heavy batteries around. On the other hand, the period for which the vehicle can be operated only on batteries should be as long as possible to reduce the overall emissions and to meet likely future requirements in city driving for operation to be only on batteries. It is this balance between battery size, heat engine size and the algorithms used in their control that determines the success of any particular hybrid combination in meeting the operational objectives. It is not surprising therefore that there are as many different combinations of battery, heat engine and control strategy as there are different HEVs.

10.1 Hybrid system configurations

We have already said at the beginning of this chapter that any vehicle that has more than one power source is classified as a hybrid, so to cover the subject fully we need to consider all the system configurations possible within the full range of HEVs. Strictly speaking, this includes all-electric vehicles using two different types of battery, or a fuel-cell and a battery, or a battery and a supercapacitor. It also includes electric vehicles in which the main battery electric drive is supplemented in peak-load conditions by power stored in a flywheel or in a hydraulic accumulator. Finally, it includes as its main category the hybrid already described in which a battery is used together with a heat engine to provide high efficiency drive combined with unlimited range.

10.2 All-electric hybrid vehicles

This class of all-electric hybrids falls for all practical, functional and design purposes into the 'pure' electric vehicle rather than the hybrid vehicle category. This is because its propulsion energy is all stored or generated electrically and the power from the main and auxiliary electrical sources used is managed and combined by entirely electrical/electronic means. Since, however, electric vehicles having two power sources have been given the hybrid label, their operation will be described in this section. Where they are discussed and described in other chapters they will be considered as particular variations on the 'pure' electric vehicle theme.

All-electric hybrid vehicles use special batteries or supercapacitors as the secondary power source in addition to the main battery. These secondary power sources are designed to provide high power for short periods under peak operating conditions – for example, during hill climbing or acceleration. This is necessary because some of the batteries with the best energy density have low power density. An extreme example of this is the aluminium-air battery, described in detail in Chapter 5. This battery has a very high-energy density of 220 Wh/kg, giving a potential vehicle operating range with a practically sized battery of 300 to 450 km (180 to 280 miles), but a power density of only 30 W/kg. Since at least 150 W/kg energy density is required for good acceleration and hill climbing performance, an auxiliary high power density source is essential. This power density is easily obtainable from a lead-acid battery and this is therefore a very suitable auxiliary battery (sometimes called a load-levelling battery) to use with an aluminium-air battery in an all-electric hybrid.

Another all-electric hybrid combination, which is under development and of increasing interest because of the improvements in fuel-cells, is the fuel-cell-driven electric vehicle with an auxiliary battery. This battery can provide the high current required for starting and can also serve as a load-levelling device which permits the fuel-cell to operate at low power initially and then warm up to operate at a steady output. This arrangement enhances the overall system efficiency and also allows the vehicle to use regenerative braking. Examples of prototype and experimental cars using this technology are given in Chapter 11.

If really high acceleration is required then the supercapacitor described in Chapter 6 could be considered as the auxiliary power source. Currently available supercapacitors, although having an energy density of only 15 Wh/kg, are capable of extremely high power densities up to 1 kW/kg, with 4 kW/kg being a development target to be achieved within three years. They are also able to accept much higher rates of charge than chemical batteries so that the high recharging rates existing under some regenerative braking conditions can be more effectively accommodated than in the case of chemical batteries. Unfortunately, supercapacitors are still only at the development stage and are therefore very expensive and unlikely to be used in production hybrid electric vehicles for some years.

10.3 Electromechanical hybrid vehicles

A further class of HEV is one in which the main drive battery is supplemented by a mechanical energy storage device such as a flywheel or a hydraulic accumulator. The arrangement of the major components for each of these alternatives is shown in Figures 10.1 (a) and (b), with the flywheel or hydraulic accumulator and pump/motor being coupled into the main power line from the propulsion batteries to the main drive motor through a central controller. This controller controls the flow of power to and from the battery, the main drive motor and the flywheel or hydraulic accumulator. The controller operates with an algorithm which optimises energy flow for maximum efficiency so that during constant speed power is only taken from the propulsion battery, but when acceleration or hill climbing is required,

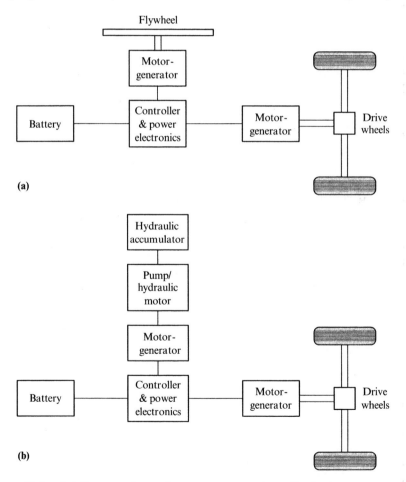

Figure 10.1   (a) Regeneration and energy storage using a flywheel
*   (b) Regeneration and energy storage using a hydraulic accumulator*

power from the flywheel or hydraulic accumulator is added to maintain performance. Conversely, when braking, the main drive motor operates as a generator and this regenerative energy is used both to recharge the propulsion battery and to speed up the flywheel or recharge the hydraulic accumulator.

Both flywheel and hydraulic accumulator are capable of supplying, or absorbing during regeneration, more than 500 W/kg during acceleration or braking and typically of storing up to 0.5 kWh of energy. Turnaround energy efficiency of these mechanical storage devices is high at about 98 per cent compared to 75–80 per cent for batteries, and as a result the energy recovered during braking can be as high as 15 per cent of the total energy used. However, problems exist of providing protection from disintegration of the flywheel in an accident, and this together with the possible requirement for two contra-rotating flywheels to overcome gyroscopic effects, makes the flywheel a potentially expensive solution. It is perhaps more suited to high rotational speed operation and eventual application as the main power source in an electric vehicle as discussed in Chapter 6. The hydraulic accumulator requires a pressure vessel in which a highly deformable membrane separates high-pressure oil pumped into it by the pump/ hydraulic motor from a compressible gas. This also requires protection to avoid any risk of failure in crash conditions. It is, however, potentially a cheaper solution for auxiliary energy storage in a hybrid of this type than the flywheel and has no vehicle stability problems.

10.4 Heat engine-electric hybrid vehicles

The hybrid vehicle on which most development work has been done to date is the one that couples a heat engine with an electric drive system. The objective remains the same as it was in 1900: to extend the range and overcome the problem of a long recharging time. It is a halfway house between the conventional internal combustion engined vehicle and the 'pure' electric vehicle, and is considered by its proponents to offer the best of both systems.

The possible combinations of heat engine and electric drive are almost infinite. At one extreme a large heat engine may be used that is capable of propelling the vehicle under most operating conditions, with an auxiliary electric drive used only for supplementing power output under high acceleration or hill climb conditions (a 'motor assist' system). At the other extreme, an electric main drive may be used with a small heat engine auxiliary power unit (APU) to supplement output and recharge the battery (a 'range extender' system). There are also two major configurations in which the components of a hybrid system can be arranged. These are the series hybrid in which transmission of power from both heat engine and electric drive to the drive wheels is primarily electrical (see Figure 10.2), and the parallel hybrid in which the transmission of power from the two sources is primarily mechanical (see Figure 10.3).

10.4.1 Series hybrids

In the series hybrid the mechanical output of the heat engine is used to generate electrical power by means of an alternator-rectifier arrangement similar to that used in a conventional car. The only difference is that this alternator-rectifier must be capable of generating sufficient electrical power to propel the car. This mechanically generated electrical power is combined with the power from the battery in an electronic controller. This controller then compares the driver demand with the drive-wheel speed and the output torque available from the main drive motor and uses its algorithm to determine the amount of power to be used from each source to meet the driver demand. This analysis takes into account all the relevant operating parameters including efficiency, performance, emissions, battery status and any special operating conditions – for example, if the vehicle is required to run only on battery power in urban areas. The controller also switches the power electronics/motor-generator to regenerative mode when the driver demands braking and directs the power being regenerated to the battery.

There are a number of advantages made possible by the use of the arrangement shown in Figure 10.2. It is possible to run the heat engine (usually an internal combustion engine) at close to constant speed and share its electrical output between charging the battery and supplying the power to drive the wheels. This minimises emissions. It is also possible to position the engine anywhere in the car outside of the passenger compartment since it does not require any mechanical connection beyond the alternator-rectifier. However, the series hybrid also has a number of disadvantages. It requires an alternator-rectifier not required by the parallel hybrid. There is a possible noise problem from the constant speed engine operating when the

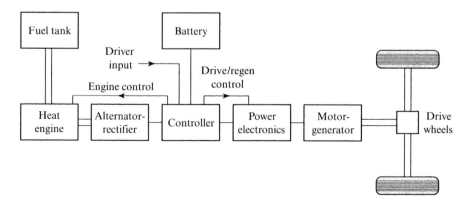

Figure 10.2   Series hybrid powertrain

vehicle is stationary. Also, the efficiency of the total system is reduced by the double conversion of mechanical power to electrical, storage of a proportion of this power in a battery and then conversion back again from electrical to mechanical power to drive the wheels.

10.4.2 Parallel hybrid

In the parallel hybrid configuration shown in Figure 10.3, the mechanical output of the heat engine is transmitted directly through a continuously variable transmission (CVT) and clutch to a three-way gearbox. In this gearbox the drive can be mechanically combined with a second drive from an electric motor energised by the battery via the power electronics. The combined mechanical output is then fed to the drive wheels. As in the series hybrid the controller compares the driver demand with the wheel speed and output torque and determines the amount of power to be used from each source to meet the driver demand consistent with obtaining best efficiency, performance and emissions. The three-way gearbox is capable of transmitting mechanical power in either direction so that during braking the clutch between the CVT and the gearbox can be disengaged and all the regenerated power can be directed to the motor-generator operating in generating mode to recharge the battery.

Instead of the CVT a stepped gearbox may be used, but it is then necessary to match the speed of the two mechanical drives into the three-way gearbox by adjusting the speed of the motor-generator during

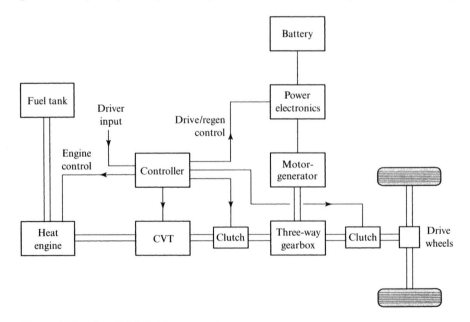

Figure 10.3   Parallel hybrid powertrain

gear-changing. With the CVT this matching can be accomplished smoothly by adjusting the output shaft speed.

One special parallel configuration that can be used is one in which the electric motor is mounted directly on to the internal combustion engine output shaft, as is done in the Honda Insight. This is really only suitable for a hybrid system in which the electric motor provides a relatively small amount of additional drive power during acceleration or hill climbing (a 'motor assist' system), although it does undoubtedly offer a neat and compact integrated drive. This arrangement does not, however, have the flexibility obtained when the system is configured in the way shown in Figure 10.3.

There are a number of advantages that a parallel hybrid configured as shown in Figure 10.3 has over the series hybrid. One is that unless the battery is low in charge, when it will be necessary for the engine to drive the motor-generator directly in regenerative braking mode, it should only be necessary to operate the engine when the vehicle is moving, so reducing the apparent engine noise problem. Another is that since only one conversion between electrical and mechanical power is made, efficiency will be much better than in the series hybrid in which two conversions are required. Finally, in the parallel hybrid case the alternator-rectifier driven by the engine in the series hybrid is not required although that cost and weight saving is largely negated by the need for additional gearboxes and clutches. The disadvantages of the parallel hybrid are that the engine can only be mounted in a few well-defined positions if the drive is to be mechanically coupled into the powertrain, and since engine speed must vary more than in the series hybrid case it is more difficult to obtain low emissions.

To obtain a measure of the improvement in efficiency and in fuel consumption possible using hybrid technology, both simulations and actual tests have been done. In [1] a computer model comparison was made between a standard 1.3 L Ford Fiesta and a parallel hybrid with equivalent body shell, driveline and weight. The arrangement of the hybrid was as shown in Figure 10.4, where either or both power sources can propel the vehicle. The components were sized according to the following rules: the internal combustion engine was chosen to allow the continuous cruising speed to be maintained without any contribution from the electric drive and with a small power surplus. The electric motor-generator was sized for acceleration and hill climbing and was about 70 per cent of the continuous power rating of the internal combustion engine. The battery capacity was chosen to provide sufficient energy to cater for the longest period of zero-emission likely to be encountered.

The results of the simulation are interesting in the three different operating conditions studied. The first condition was with both vehicles travelling at a constant 90 km/h (56 mph) for 269 s during which the internal combustion engine ran for 66 per cent of the time and developed 25 kW, of which 13 kW was used to power the vehicle and 9 kW was used to recharge the battery. The battery then powered the vehicle for the remaining 34 per cent

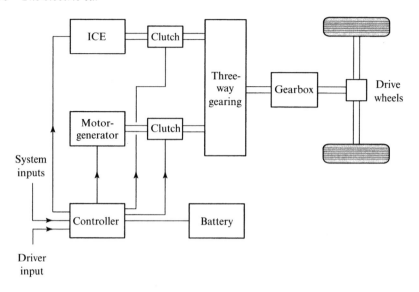

Figure 10.4   Parallel hybrid configuration used for computer simulation (Reference 1)

of the time. The resulting energy efficiency was 28 per cent for the standard Fiesta and 25 per cent for the hybrid. However, fuel consumption was improved for the hybrid with the standard Fiesta achieving 51.6 miles per UK gallon (5.47 litres/100 km), and the hybrid 59.0 miles per UK gallon (4.78 litres/100 km).

The second condition was that required by the ECE 15 urban cycle. This lasts 200 s, consists of three consecutive cycles of increasing speed and includes engine idling for 30 per cent of the time. Electric drive is used throughout the cycle except for 12 s at 50 km/h (31 mph) and a further 4.4 s during deceleration during which the engine is run on full throttle to recharge the battery. Over this cycle the energy efficiency was 10.4 per cent for the standard Fiesta and 15.6 per cent for the hybrid. Fuel consumption was 34.0 miles per UK gallon (8.20 litres/100 km) for the standard vehicle and 51.7 miles per UK gallon (5.46 litres/100 km) for the hybrid.

Finally, to simulate a typical urban traffic jam, the first part of the ECE 15 cycle was used with a longer stationary period before the cycle repeats. The authors felt that this was representative of a long queue at traffic lights. The cycle consisted of a uniform acceleration from 0 to 15 km/h (9 mph) for 8 s, followed by uniform deceleration to rest over 4 s and an idle period of 14 s. Ten of these cycles were combined to give a total of 300 s. Over this cycle the energy efficiency was 4.8 per cent for the standard Fiesta compared to 22 per cent for the hybrid. Fuel consumption was 17 miles per UK gallon (17 litres/100 km) for the standard vehicle and 63 miles per UK gallon (4.5 litres/100 km) for the hybrid.

In this simulation no attempt was made to allow for any energy recovery from regenerative braking. If this feature were included it might be expected to improve the fuel consumption figures of the hybrid by up to 10 per cent.

These three examples show dramatically how the fuel consumption of a conventional vehicle deteriorates, with the accompanying increase in emissions under heavy traffic conditions, and how much a hybrid vehicle can improve this situation.

As a comparison to the above simulation the recently (November 1999) announced Honda Insight gasoline-electric hybrid has an estimated fuel consumption of 61 miles per US gallon (73 miles per UK gallon) in city driving and 70 miles per US gallon (84 miles per UK gallon) in highway driving. In comparison, the Toyota Prius gasoline-electric hybrid has shown under actual city driving conditions a fuel consumption of 'close to 60 miles per (US) gallon (72 miles per UK gallon)' [2].

10.5 Hybrid concepts

Because of the complex combinations of main drive and auxiliary drive that are possible in an HEV, an attempt has been made in Table 10.1 to show all the practical main and auxiliary drive technologies and how they can be combined in a viable hybrid system. These main and auxiliary drive technologies have been divided into those with either a mechanical or electrical output. The flywheel falls into the electrical category because with its high rotational speed the stored rotational energy can only be accessed through an electric motor-generator. This means that the way it feeds energy into the system, or takes energy for recharging, is seen by the system as an electrical input or output. Similarly in the case of the hydraulic accumulator, although it is not impossible to take a mechanical drive from the hydraulic motor driven by the accumulator, it is far more convenient to convert it into an electrical form.

Table 10.1 shows that there are many possible combinations of main and auxiliary drive, some being more practical than others. The more probable combinations are marked A with the possible but unlikely ones marked B. If we consider only the probable combinations it can be seen that there are 11 of these. The number of options is further increased since those having a heat engine as the main drive can be operated in either the series or parallel configuration. This adds three further options and makes a total of 14. Also, in the parallel configuration there are a number of different ways of transmitting the mechanical drive to the wheels. The three most important of these are shown in simplified form in Figure 10.5.

The first drivetrain shown (Figure 10.5(a)) uses a common gearbox and corresponds to that shown in Figure 10.3. This is the most commonly used in current experimental HEVs, although the second drivetrain shown in Figure 10.5(b), in which the electric motor is mounted on the same shaft as the heat engine, is becoming much more frequently used, for example in the

Table 10.1   Combinations of main and auxiliary hybrid drive technologies

		Main drive					
		IC engine	Gas turbine	Stirling engine	Battery-electric motor	Fuel cell-electric motor	Flywheel-electric motor
Auxiliary drive	IC engine	–	–	–	A	B	A
	Gas turbine	–	–	–	A	B	B
	Battery-electric motor	A	A	A	A	A	A
	Flywheel-electric motor	B	B	B	A	B	–
	Super-capacitor	B	B	B	A	B	–
	Hydraulic accumulator-electric motor	B	B	B	A	B	–

A = Probable combinations
B = Possible but unlikely combinations

recently announced Honda Insight. The third option shown in Figure 10.5(c) requires the use of two separate drive outputs, one mechanical and one electrical, to the front and rear wheels of the vehicle, respectively. To the author's knowledge this precise arrangement had only been used in one experimental hybrid, the Audi Duo (now out of production), until the recent announcement (October 2000) of the DaimlerChrysler Dodge Durango hybrid that is scheduled to be on the market in 2003 (see Table 10.3). However, it is of more than academic interest since it provides a low-cost and flexible way of connecting the two drives through the road wheel/ surface interaction at opposite ends of the vehicle. Interestingly Toyota has recently announced [3] a prototype minivan which is a variant of this system in which the front wheels are driven by an ICE in a parallel configuration with an electric motor, while the rear wheels are driven by a separate battery-electric motor drive. Renault has also announced two similar proto- type split drive cars: the Next and Koleos. All three of these are also listed in Table 10.3. It is clear that the coordination and optimisation of the separate mechanical and electrical drives to the front and rear wheels of these vehi- cles requires sophisticated electronic controls if smooth operation is to be achieved. There is also the disadvantage that those variants where the mechanical drive and electrical drive are on separate axles require the axle with the mechanical drive to generate additional power and to drive the axle with the electrical drive into regenerative mode if the battery is to be recharged during operation.

The first drivetrain option (Figure 10.5(a)) using a common gearbox has already been taken into account in calculating the number of hybrid options

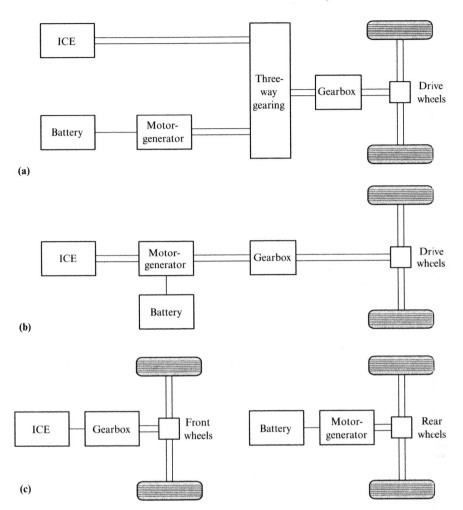

Figure 10.5 (a) Separate drive inputs, (b) common drive shaft and (c) separate drives to front and rear wheels

already listed. The other two drivetrain options (Figures 10.5(b) and (c)), either of which could be used with any of the five A class heat engine/ battery-electric motor combinations shown in Table 10.1, gives a further ten possible options, making a total of 24 hybrid options in all. In addition to this, the main drive and auxiliary drive can vary in relative power over a wide range. It is not surprising therefore that every HEV prototype announced has a different combination of features and technology.

Interest in hybrids has increased recently since the announcement in the late 1990s of production hybrids from Honda with the Insight, and Toyota with the Prius. Also, Nissan are reported to be about to launch a production

hybrid called the Tino. Both the Honda and Toyota vehicles are on sale for
$18 000 in the USA, although the true production cost is believed to be
about twice this figure. This subsidised price should help to ensure that there
are significant numbers of hybrid vehicles in the hands of the USA public
within the next two years and great interest will be taken in how they
perform in everyday use. However, until the production cost can be reduced
to a level at which the manufacturer can make a profit, which is a significant
challenge with effectively two propulsion systems on each vehicle, it is difficult to see how large numbers of hybrid vehicles can be sold and the environmental advantages of using them realised.

10.6 Production hybrid cars

To complete this chapter, details of the three hybrid electric cars developed
by major manufacturers and currently (2000) in production and available
for purchase by the general public are given in Table 10.2 and short descriptions of the Honda and Toyota hybrids are given below. Insufficient data is
available at the time of writing to give detailed information on the Nissan
Tino. Prototype and experimental hybrid cars are considered in Section
10.7. The technical details given have been provided directly by the manufacturers except in a few cases where information has been obtained from
published press reports. In the case of the prototype and experimental cars,
details inevitably are incomplete but nevertheless give a good idea of the
current status of the technology and the number of manufacturers involved.
Fuel-cell cars, although strictly speaking also hybrids because of their use of
two different electrical sources (fuel-cell plus battery), are best considered as
a separate class of electric vehicles and their technical details are given in
Chapter 11.

The Honda Insight and the Toyota Prius share the distinction of being
the first two production hybrid electric cars to be offered for sale to the
general public, the availability of the Nissan Tino being only announced in
April 2000. The Insight and Prius have been on sale in Japan since 1998,
with 35 000 Prius vehicles reported as having been sold up to the end of
1999. They are both now available in the USA and Europe from the early
months of 2000 at a subsidised price. The essential technical and performance details for these two vehicles are given in Table 10.2.

10.6.1 The Honda Insight

The two-seater Honda Insight uses an integrated motor assist (IMA) hybrid
system (see Figure 10.6) that features a high-efficiency gasoline engine, electric motor and lightweight five-speed transmission. This is used in combination with a lightweight, aerodynamically efficient ($C_d = 0.25$) aluminium and
plastic body to obtain an excellent on-road performance of 0 to 100 km/h

Table 10.2   Hybrid production cars

Manufacturer	Honda	Toyota	Nissan
Model name	Insight	Prius	Tino
Hybrid type	Parallel (motor assist)	Parallel	
Heat engine	Lean burn gasoline VTEC IC	Atkinson cycle gasoline IC with VVT	4-cyl gasoline IC
Capacity (cc)	995	1 500	1 800
Max power O/P (kW) (heat engine)	50	43	
Electric drive	PM brushless DC	PM synchronous AC	PM synchronous AC
Max power O/P (kW) (electric drive)	10	30	17
Battery type	NiMH	NiMH	Li-Ion
System voltage (V)	144	288	
Battery energy capacity kWh	0.94	1.9	
Battery-charging method	From regen. braking only	From ICE & regen. braking	
Transmission	5-speed manual or CVT	Torque split (electrical CVT)	
Max speed (km/h)	180	160	
Av. fuel cons. (km/litre*) (litres/100 km)	28 (3.4)	29 (3.45)	
Sale price $	18 000	18 000	28 500
Production cost $ (est.)	28 000	35 000	31 000

* 1 km/litre = 2.82 miles/UK gallon = 2.35 miles/US gallon

(62 mph) in 12 s with a top speed of 180 km/h (112 mph) and a fuel consumption of only 28 km/litre (83 miles per UK gallon) on the EU Combined
Cycle test. The three-cylinder 995cc gasoline engine uses SOHC VTEC lean-
burn technology combined with low friction components and lightweight
materials to give what is claimed to be the world's lightest engine. Already
low emissions are further improved by the use of a new lean-burn compatible NO_x catalyst.

Electric-motor assistance is provided by an ultra-thin (60 mm) DC
brushless motor (see Figure 10.7) directly connected to the crankshaft, this

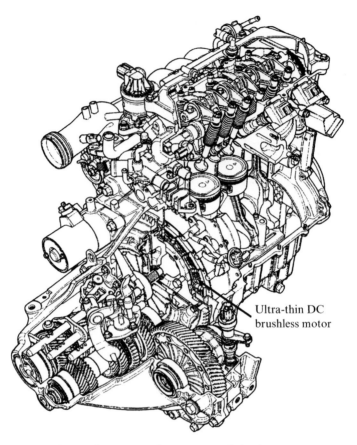

Ultra-thin DC
brushless motor

Figure 10.6   Honda Insight integrated motor assist hybrid system

motor being powered through an advanced electronic power control unit from a 144 V nickel-metal hydride (NiMH) battery with a power capacity of 0.94 kWh and weighing 20 kg. This battery is charged by regenerative braking alone, but this is apparently sufficient to maintain charge under normal operating conditions and allow the electric motor to be used to boost engine performance during acceleration and hill climbing to the level of a 1.5-litre gasoline engine. Engine output being increased from 50 kW to 56 kW with electric-motor assistance and torque from 91 Nm at 4 800 rpm to 113 Nm at 1 500 rpm. A photograph of the Insight vehicle is shown in Figure 10.8.

10.6.2 The Toyota Prius

The five-seater Toyota Prius hybrid operates with more equal sharing of the power between the gasoline heat engine and the electric motor than is the

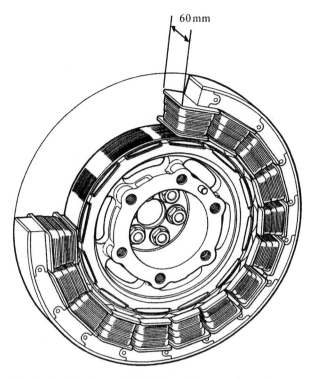

Figure 10.7   Honda Insight ultra-thin DC brushless motor

Figure 10.8   Honda Insight hybrid electric car

case for the Honda Insight. This is done by means of a hybrid transmission with a power-splitting device using a planetary gear system (see Figure 10.9). The engine drive shaft is connected to the planetary gear carrier, which allows power to be simultaneously supplied through the outer ring gear to the wheels, and through the sun gear to the generator. The electricity produced by the generator can then be directed to the electric motor to increase the power available to drive the car or through the inverter to be converted into direct current to charge the battery. By controlling generator speed, the hybrid transmission acts as a continuously variable transmission that can freely vary the engine speed.

The way in which the Toyota hybrid system operates under a number of different operating conditions is shown in Figure 10.10. These are as follows:

1. When starting out, moving at very low speeds, descending long, gentle slopes, and at other times when running the engine would be inefficient, the engine is turned off and the car is propelled by the electric motor only (A).
2. During normal operation, the engine's power is split, with part going to drive the vehicle (B) and part being used to generate electricity. This goes to the motor which then assists in powering the car (C).
3. During full-throttle acceleration, extra energy is drawn from the battery for additional motive power (A).

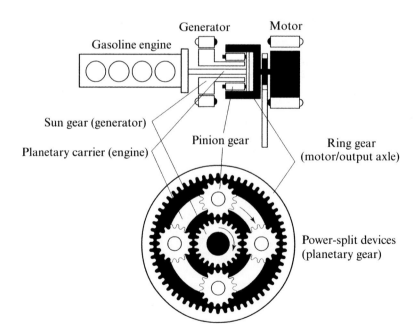

Figure 10.9   Toyota Prius power-splitting device using a planetary gear system

4. During deceleration or braking, the motor acts as a generator, transforming the kinetic energy of the wheels into electrical power that is then converted to DC by the inverter and then stored in the battery (A).
5. If the battery state of charge is low, the control system automatically sends more of the engine power to generate electricity and recharge the battery (D).

When the car is at rest, the engine is stopped. Hydraulic brakes are used as a back-up for the regenerative braking system although regeneration always takes priority.

Special high-performance NiMH batteries of 1.9 kWh capacity are used in the Prius. These are claimed to have three times the power output of the

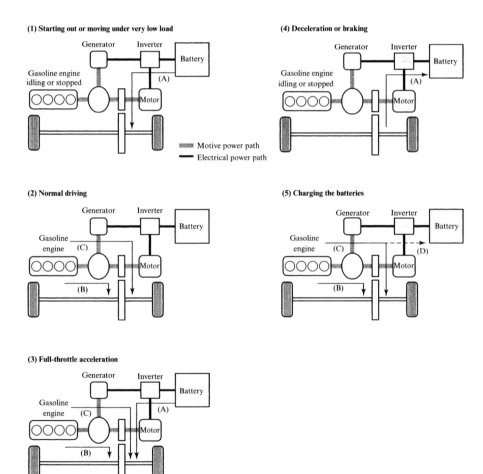

Figure 10.10   Toyota Prius operating conditions

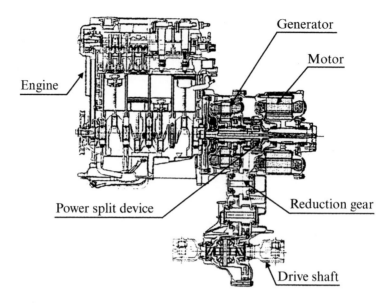

Figure 10.11   Toyota Prius engine and hybrid transmission system

NiMH batteries normally used in electric vehicles and in common with most hybrid electric vehicles they require no external charging since their charge is maintained by the internal combustion engine driven generator and by regeneration.

Figure 10.12   Toyota Prius hybrid electric car

Table 10.3   Prototype and experimental hybrid cars

Manufacturer	BMW	BMW	Citroen	Citroen	Daimler Chrysler	Daimler Chrysler	Fiat	Ford	Ford	GM
Model name	518 Hybrid	3 Series Hybrid	Xsara Dyn-active	Saxo Dynavolt	ESX 3	Dodge Durango TTR (SUV)	Multipla	P2000 Prodigy	Escape HEV (SUV)	Precept
Hybrid type	Parallel	Series	Parallel	Parallel (range extender)	Parallel (motor assist)	Parallel split drive front-elect. rear-ICE	Parallel	Parallel (motor assist)	Parallel (motor assist)	Parallel split drive
Heat engine	4-cylinder IC	4-cylinder IC	4-cylinder IC	2-cylinder IC	DI diesel	V6 gasoline IC	4-cylinder IC	DI diesel	Atkinson cycle 4-cyl gasoline IC	DI diesel
Capacity (cc)	1 600	1 600	1 360	400	1 500	3 800	1 600 (16 V)	1 200	2 000	1 300
Max power O/P (kW) (heat engine)			55	6.5	54	52	76	55		40
Electric drive	3-phase induction	PM synch		Separately excited DC	AC induction	3-phase induction	3-phase induction	3-phase induction	PM	2×3 phase induction
Max power O/P (kW) (electric drive)	26		25	20	15		30	35	65	25 front 10 rear
Battery type	NiMH		NiCd	NiCd	Li-Ion		NiMH	NiMH	NiMH	Li-Ion or NiMH
System voltage (V)	200		168	120	300		216		300	330
Battery energy capacity (kWh)							19			3
Battery-charging method	From ICE & regen.		From ICE & regen.	From ICE & mains supply		From regen. braking	From ICE & mains	From ICE & regen.	From regen. braking	From ICE & regen.
Transmission					Electro-mech. auto	Electronic-controlled manual	4-speed mech/auto	5-speed elect. stepped		
Max speed (km/h)			170	120			160			
Av. fuel cons.* (km/litre) (litres/100 km)			19 (5.2)		30.5 (3.3)	7.9 (12.6)	14.7 (6.8)	34 (2.94)	17 (5.8)	34 (2.94)
Date for production						2003	1 year after Multipla IC prod. car	2003	2003	2005+

* 1 km/litre = 2.82 miles/UK gallon = 2.35 miles/USA gallon

Table 10.3 (cont'd.)　Prototype and experimental hybrid cars

	GM	GM	Mitsubishi	Mitsubishi	Nissan	Renault	Renault	Renault	Renault	Toyota	Volvo
Manufacturer Model name	EV1 Hybrid	Silverado (SUV)	HEV	ESR	Neo Hybrid	Next	Vert	Modus	Koleos	HV-M4 Minivan	ECC
Hybrid type				Series	Parallel (motor assist)	Parallel split drive	Dual motors	Dual motors	Parallel split drive	Parallel split drive	Series
Heat engine	Diesel	V6 gasoline	ICDI CNG	IC gasoline	IC gasoline	IC	Diesel turbo alternator	Turbo alternator	Gasoline turbo IC	Lean burn gasoline	Diesel gas turbine
Capacity (cc)		3 600	1 500		1 800	750			2 000	2 400	
Max power O/P (kW) (heat engine)		164	20			35	38	38	126		
Electric drive		2×3 phase induction	2×3 phase induction		PM Synch	Dual PM Synch in wheels	Dual PM Synch	Dual PM Synch			
Max power O/P (kW) (electric drive)		24	60			14	90	90	30		
Battery type		Li-Ion	Li-Ion		Li-Ion	NiCd	NiCd	NiCd	Li-Ion		NiMH
System voltage (V)			336					300			
Battery energy capacity (kWh)							6.6	55			
Battery-charging method			From ICE & regen.				From turbo				
Transmission										CVT	
Max speed (km/h)						167	165				
Av. fuel cons.* (km/litre) (litres/100 km)	34 (2.94)	14.8 (6.7)				29 (3.4)					
Date for production		2004	2001								

* 1 km/litre = 2.82 miles/UK gallon = 2.35 miles/USA gallon

Toyota has specially developed the 1.5-litre gasoline engine for use in hybrid systems. It uses the Atkinson high-expansion cycle for improved efficiency. This makes it possible to use very small combustion chambers that give a high expansion ratio and increase the amount of combustion energy that is converted to useful work. The engine's low revolution rate (4 000 rpm max.), permits the use of lighter-than-normal moving parts, a smaller diameter crankshaft, lighter piston-ring springing, lower valve-spring loads and significant reduction in internal friction. Variable valve timing (VVT) is also used to continuously match valve timing to engine operating conditions, resulting in a lightweight engine of excellent performance and superior fuel efficiency. The full engine and hybrid transmission assembly is shown in Figure 10.11 and a photograph of the Prius car is shown in Figure 10.12.

10.6.3 The Nissan Tino

Although little information is available so far on the Nissan Tino, it is of particular interest in the author's opinion because of Nissan's use of a lithium-ion battery to power its electric drive. This may be the forerunner of much more extensive use of this very promising battery in production electric vehicles.

10.7 Prototype and experimental hybrid cars

The information available on prototype and experimental hybrids is listed in Table 10.3. Of the 21 vehicles listed, only six have been given production dates by their manufacturers, and on a further five the manufacturers have released very limited information. However, the number of cars being developed indicates the strong interest in the hybrid as a practical and flexible engineering solution to the range and refuelling problems of the 'pure' electric car.

Of particular note is the use in the GM Precept, the Renault Next and Koleos, and the Toyota HV-M4 minivan, of a split drive system using a combined heat engine/electric drive to the front wheels together with a separate electric drive to the rear wheels. Also of considerable interest is the DaimlerChrysler Dodge Durango in which the front wheels are driven only by the electric motor and the rear wheels only by the internal combustion engine. This arrangement is almost identical to the now-discontinued Audi Duo hybrid in which the internal combustion engine drove only the front wheels and the electric motor drove only the rear. In hybrid systems of this type it is possible to dispense with the complex and expensive gear box required to combine the electrical and mechanical drives when both drive the same axle and instead use the road as the energy transmission route between the two drives. It will be interesting to see if in use on the road tyre wear is significantly increased.

As might be expected all the electric drive motors used are AC and all models currently under development use NiCd, NiMH or Li-ion batteries.

The major inhibiting factor in making hybrid cars acceptable to the consumer remains the high cost of a car with two separate propulsion systems. It is difficult to see how this can be overcome in view of the actual production costs (see Table 10.2) of the Honda and Toyota production hybrid cars now being sold to the public at a heavily subsidised price.

References

1   NEWMAN, P. and MILNER, P.: 'Minimising environmental impact, maximising marketability, the HDT approach', Electric Vehicle Seminar, Motor Industry Research Association (MIRA), Nuneaton, Warwickshire, England, April 1992
2   MOORE, W.: 'EV world update', November 14, 1999, http://www.evworld.com
3   MOORE, W. 'EV world showcases HVM4 hybrid-electric minivan', December 1999, http://www.evworld.com

General References

1   WAKEFIELD, E. H.: 'History of the electric automobile – hybrid electric vehicles' (SAE, 1998)
2   'Hybrid electric vehicles', 11 papers from the 2000 SAE Future Transportation Technology Conference (SAE, 2000)

Chapter 11

Fuel-cell electric cars

In Chapter 6 the use of the proton-exchange membrane (PEM) fuel-cell as an energy source suitable for an electric car was discussed and its method of operation described. In this chapter we will consider how the hydrogen essential for the operation of a PEM fuel-cell can be provided in the car, how it can be stored, and what the implications for the fuel infrastructure will be. All the information currently available on the prototype and experimental fuel-cell cars being developed by the major automotive manufacturers is also given in Table 11.1.

11.1 Hydrogen fuelling

When PEM fuel-cells are used in a vehicle, hydrogen fuel is required. If hydrogen is to be used as the prime and only fuel then it can be stored in an insulated tank in cryogenically cooled liquid form, compressed and stored in a pressure tank, or dispersed in metal hydrides or possibly in carbon compounds. The simplest and least expensive method is to compress the hydrogen and store it in a pressure tank made of stainless steel or aluminium. However, it requires a considerable amount of energy to compress the hydrogen to the 400 or so atmospheres required. To compress hydrogen to 400 atmos requires between 20 and 40 per cent of the energy content of the hydrogen being compressed [1], although since this energy would normally be expended in the hydrogen supply chain rather than on the vehicle, the efficiency of the vehicle itself is not affected. This method also requires a heavy pressure tank that occupies a substantial amount of space in the vehicle. The alternative use of a heavily insulated tank of liquid hydrogen at –253 °C overcomes the high storage weight problem but brings with it its own problems affecting efficiency, which include the energy required to cool the hydrogen in the first place and to keep it cool on the vehicle. There is also a significant loss of hydrogen boiling off during refuelling and a small loss during on-board storage. Whichever of these two methods is used, the volume of the tank in which the hydrogen is stored needs to be at least four times that of a full tank of gasoline for an equivalent distance of travel.

An alternative to the above methods is to use metal hydride beds in which hydrogen can be bonded with a metallic compound at a low temperature and released again when the temperature is raised. This makes it possible to limit the rate at which hydrogen can be extracted and therefore to reduce the risk of explosion. This is an issue of some importance where significant quantities of hydrogen are stored and where, as in a vehicle installation, there is a risk of a crash resulting in rupture of the storage system. However, it is difficult to store sufficient hydrogen by this method to give the vehicle range required. An alternative solution that may overcome some of these problems is the use of graphite nano-fibres (carbon whiskers) for hydrogen storage. These fibres, which are under investigation at a number of research laboratories, act as an absorbant with the property of being able to store large amounts of hydrogen in a relatively small volume at low to moderate pressure. Considerable development to make this storage system practical is still required and even if this development is successful it still does not overcome the potential safety issues raised by the need to frequently refill the vehicle tank with hydrogen at recharging stations.

11.2 Reforming

Many of these problems can be overcome if the hydrogen is generated on-board at normal pressure and temperature as required and a process known as 'reforming' can achieve this. This process involves steam reforming or partial oxidisation of a suitable liquid hydrocarbon or alcohol fuel such as methanol, ethanol, natural gas or gasoline. Steam reforming, in which steam reacts with the hydrocarbon/alcohol fuel in an endothermic reaction to produce hydrogen, has the highest fuel efficiency, producing the maximum amount of hydrogen for a given amount of fuel. It has usually been the method used to obtain hydrogen from methanol. The partial oxidation method where the hydrocarbon or alcohol reacts with oxygen in an exothermic reaction to produce hydrogen has been used to obtain hydrogen from gasoline. A disadvantage of the reforming process is that it requires a few minutes for the reformer to warm up when first switched on and in vehicle use this must be accommodated by the use of a supplementary battery to provide power for the vehicle during this initial period.

Johnson Matthey have developed a methanol reformer called the 'Hot Spot', illustrated in Figure 11.1, which combines the advantages of both types of reformer [2, 3]. It uses heat produced by the partial oxidation process to drive the heat-absorbing steam-reforming reaction, and makes possible a physically compact reformer unit that has fast start-up and good fuel efficiency. A single module, which is about the size of a soft drinks can, is capable of generating more than 750 litres of hydrogen per hour, sufficient to power a 750 W fuel-cell. This reformer is being developed to operate with a range of other fuels, with gasoline being of particular interest because of its ready availability through the existing distribution infrastructure. A sophisticated multistage catalytic filtering process is required to reduce the 20–25 per cent of carbon dioxide and the 1–3 per cent of carbon monoxide present in the hydrogen gas output from the reformer. This avoids poisoning of the platinum anode catalyst in the fuel-cell and reduces overall vehicle emissions. Small concentrations of carbon monoxide will cause severe anode catalyst poisoning in a PEM fuel-cell, with 20 ppm causing at least a 50 per cent loss of efficiency, while high concentrations of carbon dioxide (over 20 per cent) cause indirect poisoning of the anode catalyst [3]. The filtering process required to overcome this problem will reduce the fuel conversion efficiency since about 6 per cent of the hydrogen will be oxidised during the process, and improved catalysts which are more tolerant to this poisoning are being developed. The efficiency of conversion, allowing for parasitic losses, of methanol to electricity is about 38 per cent [4], when the further losses of DC/AC conversion and propulsion motor and gearing are added the overall efficiency from methanol fuel to the road wheels is about 30 per cent.

The ideal solution would be a fuel-cell that operates directly with hydrocarbon or alcohol fuels. This seems to be a possibility and a PEM design that runs on air plus a mixture of methanol and water (a direct methanol

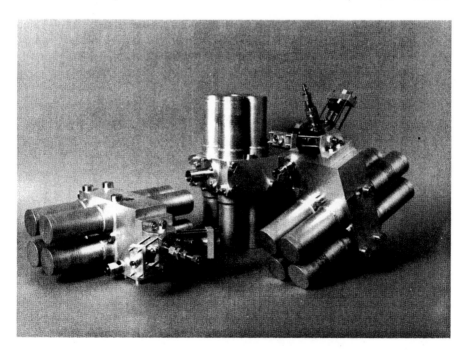

Figure 11.1   Johnson Matthey 'Hot Spot' methanol reformers

fuel-cell (DMFC)) is being considered [3]. The realisation of such a design requires better catalysts to overcome losses at the anode and improved membranes and cathode catalysts to overcome cathode poisoning by the methanol. Although it is likely that a fuel-cell of this type would only have about 25 per cent of the power density of a conventional hydrogen/air cell, the reduction in cost and complexity over a reformer/PEM fuel-cell vehicle installation may make it the preferred option.

The advances made in PEM fuel-cell technology in recent years have excited considerable interest in all the companies involved in the development of electric vehicles. DaimlerChrysler has been testing a number of fuel-cell vehicles since 1993 using a variety of fuels and in its new partnership with Ford and Ballard expects to demonstrate a commercially viable fuel-cell car during the year 2001. General Motors has shown a fuel-cell van powered by a 50 kW PEM cell stack using methanol and were reported to be intending to offer a production-ready fuel-cell vehicle by 2004. It is understood, however, that they may now replace methanol with stored hydrogen as they have reservations about methanol toxicity. Ford has demonstrated a fuel-cell-powered passenger car, the P2000 (see Figure 11.2), using stored hydrogen and is reported to have also demonstrated a sport utility version of this (the Ford Escape SUV) using a fuel-cell/methanol reformer power plant.

In Europe a joint PEM fuel-cell programme to reduce weight and cost is in progress involving Peugeot/Citroen, Renault, Volkswagen and Volvo and a variety of fuel-cell and hybrid-powered vehicles are being developed. Japanese automotive companies, Toyota, Mazda, Nissan and Honda, have also been actively developing concept fuel-cell-powered cars and have plans to begin selling production versions by 2003.

hydrogen tank   fuel-cell assembly

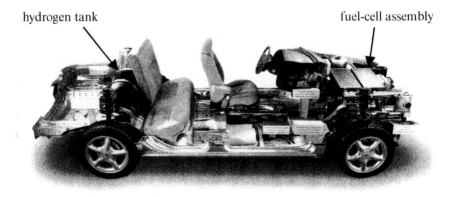

Figure 11.2   Ford P2000 fuel-cell-powered passenger car using either stored liquid or compressed hydrogen

There have been recent suggestions from BMW and Delphi Automotive that a high-temperature solid oxide fuel-cell (SOFC) may be a better candidate for vehicles in the long term. An SOFC is able to accept much lower purity hydrogen than the PEM fuel-cell and as a consequence requires a much simpler and cheaper reformer to generate the hydrogen it uses from a hydrocarbon primary fuel. It does, however, operate at a temperature of 600–1000 °C, which presents considerable operational problems, not least of which is its warm-up time of up to 45 min. To operate in a vehicle with this restriction would require a supplementary battery with sufficient capacity to run the vehicle during the warm-up period. A goal has been set by the US Department of Energy to reduce this warm-up time to less than 5 min, but this is likely to require development over a considerable period.

If the maximum efficiency is to be obtained from a fuel-cell there is no doubt that this can only be achieved by the direct use of clean hydrogen. The efficiency of a fuel-cell powertrain fed by hydrogen is currently about 40 per cent with the potential to reach 60 per cent eventually. In contrast, if a methanol reformer is used the overall efficiency of the reformer/fuel-cell powertrain drops to less than 30 per cent, similar to a gasoline/electric hybrid. The efficiency seems likely to be even worse for a fuel-cell operating with a gasoline reformer. The two main advantages of reformer/fuel-cell systems are the ability to use hydrocarbon or alcohol fuels directly, generating much lower levels of pollution than when these fuels are burnt in a conventional internal combustion engine, and the ability to make use of the existing fuel distribution and refuelling system. This should make it possible to have fuel-cell electric cars on the road in small production quantities before 2007.

An alternative view is that fuel-cells are only going to be a useful technology if they are fed by clean hydrogen carried in quantity on the vehicle. In fact one view is that in the longer term hydrogen will replace fossil fuels almost entirely and we shall have a so-called 'hydrogen economy.'

11.3 Infrastructure

If anything approaching a 'hydrogen economy' is ever to be realised, it will be necessary to be able to produce and supply hydrogen to the public in the same general way as gasoline and diesel oil. This requires the development of safe methods of generating hydrogen either centrally, at the refuelling station, or even in the home; the development of safe methods of distributing centrally generated hydrogen; and the development of effective means (manual or robotic) of connecting the high-pressure or cryogenically cooled hydrogen supply to the fuel-cell vehicle. If plenty of electricity is available that has been generated by non-polluting methods such as solar, hydroelectric or nuclear power or by wind or tides, then the centralised large-scale production of hydrogen by the electrolysis of water with distribution to and storage at refuelling stations seems likely to be preferred. An added advantage of this centralised production is the ability to make full use of the intermittent nature of many of the non-polluting energy sources. Alternatively, centralised production of hydrogen using steam-reforming of hydrocarbons may be used. In either case, distribution by pipeline is likely to be preferred to road or rail transport.

It is also possible to use either of these hydrogen production methods on a smaller scale at the refuelling station itself. Intriguingly it is also possible to envisage home-based electrolysis and compression systems producing hydrogen overnight for use in the family's fuel-cell vehicle the next day, although safety issues may make this unacceptable. Refuelling stations (and domestic installations) would require pressure tanks to store the hydrogen and foolproof methods of refilling vehicle tanks, but in due course these will be developed. It seems likely, however, that development of this 'hydrogen economy' infrastructure will take a considerable time so that in the shorter term if fuel-cells are to be used it seems more likely to be with on-board reformers fed with methanol, gasoline or natural gas.

11.4 Safety

For mainly historical reasons hydrogen is seen as highly explosive and a substantial safety hazard. It is certainly easy to ignite over a wide range of hydrogen/air mixtures and burns with an invisible flame that could mislead users. However, because of its low weight when released into the atmosphere and its rapid dissipation in air, it is probably less dangerous in the open air than the gasoline we keep stored every day under the rear seat of our cars. The major problem that could arise is when a hydrogen-fuelled vehicle is in an enclosed space such as a garage. Under these conditions a small leakage could result in the build-up of a dangerously explosive air/ hydrogen mixture which could be triggered by the slightest spark and cause a large explosion.

The other major problem is ensuring that the on-board tank in the vehicle can be filled safely with high-pressure or low-temperature hydrogen at the refuelling station. This may well need to be done by using robotic methods of connecting the vehicle to the hydrogen supply, a technology already being pioneered by BMW [1]. The cost of redeveloping refuelling stations to use this type of technology has been estimated by Ford at about four times the cost of refurbishment of a conventional filling station [5]. No doubt extensive safety testing will be required to reassure the public that both the filling methods finally developed and the on-board storage and use of hydrogen is as safe or safer than gasoline.

Table 11.1 Prototype and experimental fuel-cell electric cars

Manufacturer	Daimler Chrysler	Daimler Chrysler	Ford	Ford	GM	GM	Honda	Mazda
Model name	NECAR 5	Commander SUV	P2000 HFC	Think Focus FCV	Opel Zafira	Opel Zafira HydroGen 1	FCX-V3	Demio-FCEV
Drive type	3-phase induction		3-phase induction	3-phase induction	3-phase induction		PM Synch	3-phase induction
Power source	Fuel-cell + methanol reformer or H storage	Fuel-cell + methanol reformer + battery	Fuel-cell + H storage	Fuel-cell + H storage	Fuel-cell + methanol reformer or H storage	Fuel-cell + H storage	Fuel-cell + H storage + super-cap.	Fuel-cell + H storage + super-cap.
Max power O/P (kW)	55	70	67	67	80	89	60	65
Voltage (V)	330		255	315				
Top speed (km/h)	145		128	128	120	145	130	90
Claimed range (km)	450		160	160	640	400	177	170
Date for production	2004	2004?	2004	2004	2004		2003	

Table 11.1 (cont'd.)   Prototype and experimental fuel-cell electric cars

Manufacturer	Mitsubishi	Nissan	Peugeot/Citroen	Renault/Volvo Euro. Project	Toyota	VW	VW
Model name	Fuel-cell EV	FCV	Partner	Fever	FCEV	Bora Hymation	Sharan
Drive type		PM Synch		Synch wound rotor	PM Synch	3-phase induction	
Power source	Fuel-cell + reformer	Fuel-cell + reformer	Fuel-cell + reformer or H storage	Fuel-cell + H storage + NiMH batt.	Fuel-cell + methanol reformer	Fuel-cell + H storage	Fuel-cell + reformer
Max power O/P (kW)				30	50	89	
Voltage (V)				FC 90 system 250			
Top speed (km/h)				120	125	140	
Claimed range (km)				400	500	350	
Date for production	2005	2004/5	2003/4	2003+			

11.5 Prototype and experimental fuel-cell electric cars

All the major automotive manufacturers have been working on fuel-cell electric cars and the information available on these prototype and experimental cars is collected in Table 11.1. The table lists 15 fuel-cell cars. The hydrogen required for the fuel-cells is either obtained by methanol reformer in five cases, from direct storage on the vehicle in seven cases, or with the alternative of either source in three cases. The move from the use of a reformer to direct pressure storage of hydrogen on the vehicle has been noticeable in the experimental and prototype fuel-cell cars announced during the year 2000. This perhaps reflects the amount of development still required on reformers suitable for use in passenger cars and some easing of the concerns voiced in the past on the safety issues involved in storage and refuelling of vehicles using hydrogen. If, however, the fuel-cell electric car is to fulfil its promise as a method of directly using hydrocarbon fuels supplied through the existing infrastructure in a more efficient way than can be achieved in the internal combustion engine, it will be necessary to continue with the development of fuel-cells using either separate reformers or with the capability of running directly on hydrocarbon fuels.

Firm dates for production versions of these cars have been given by the manufacturers and are indicated in the table. If these rather optimistic dates are met, it would appear that the 2004–2005 period will really begin to show how viable fuel-cell technology is in everyday use by the general public.

References

1   PEARCE, F.: 'Kicking the habit', *New Scientist*, November 12th 2000, pp. 35-42
2   HOOGERS, G.: 'Fuel cells: power for the future', *Physics World*, August 1998, pp. 31–36
3   GRAY, P. G. and FROST, J. C.: 'Impact of catalysis on clean energy in road transportation', *Energy and Fuels*, 1998, **12**, pp. 1121–1129
4   CARPENTER, I., EDWARDS, N., ELLIS, S. *et al.*: 'On-board hydrogen generation for PEM fuel cells in automotive applications', SAE Technical Paper Series 1999-01-1320
5   'Green State', article in *Daily Telegraph* (London), 20 January 2001

Chapter 12

Economics of electric cars

If the electric vehicle and particularly the electric car is ever to compete effectively with conventional internal combustion-engined vehicles, the price at which it can be profitably built and sold to the public must be similar to that for a conventional vehicle having a similar feature level. A small premium may be acceptable for the electric vehicle providing it offers a significant reduction in emissions and in the case of fuel-cell and hybrid electric vehicles, a significant reduction in fossil-fuel consumption.

12.1 Electric car cost comparisons

At present, prices of those production electric and hybrid cars that are currently (February 2001) available, are in the region of twice that for comparable internal combustion-engined cars. Examples are shown in Table 12.1 with all prices in US dollars for easy comparison. These sale prices, with the exception of the hybrids and the subsidised Renault Clio Electric, largely reflect the cost of the batteries used. The cost of these electric cars excluding their batteries is generally less than comparable conventional cars as can be seen in the case of the Citroen AX/Saxo and Peugeot 106 where the battery cost is excluded from the purchase price. This cost is lower because of the relative simplicity of electric car powertrain design in which an electric motor with a one-step gearbox, controller, battery and charger takes the place of a complex internal combustion engine with many working parts, a catalytic converter, multiple-step or automatic gearbox, starter, alternator and gasoline tank.

A cost comparison of various electric vehicle powertrain/drivelines has been published in [1] and is shown in Table 12.2. Costs are referenced to the cost of a conventional gasoline IC, shown as 100. The only item I would take issue with the author on is the cost quoted for the hybrid IC, which I would put currently as being similar or even a little higher than the EV range extender hybrid. In the case of both these hybrid types, the cost of having two propulsion systems will continue to inhibit their sales where the full cost is passed on to the customer. In the case of the 'pure' battery EV it is the battery which constitutes the major cost, as it is the fuel-cell and its reformer in electric vehicles powered by that device. As stated in [1] the parallel hybrid powered by a DI diesel engine is likely to be the most practical way in the short term of meeting the generally accepted future fuel consumption target for hydrocarbon fuelled cars of 3 litres/100 km (33 km/litre or 80 miles/US gallon). However, developments in computer-controlled lean burn internal combustion powertrains are likely to make the '3 litre' target achievable in the longer term by many of the smaller conventional vehicles without going to the cost and complexity of the hybrid. Where the hybrid scores is by making zero-emission possible when it is run in electric-

Table 12.1   Sale and lease cost of production hybrids and electric cars

	Lease cost ($)	Sale cost ($)	Estimated production cost ($)	Battery type
HYBRIDS				
Honda Insight		18 000 (in USA)	28 000	
Toyota Prius		18 000 (in USA)	35 000	
Nissan Tino		31 000 (in Japan)	31 000+	
ELECTRIC CARS				
Citroen AX/Saxo		12 300 excluding battery		NiCd battery required
Daihatsu HiJet EV		23 990		
GEM EV	35 per hour	10 000		?
GM EV1	399–480 per month	33 995		Lead-acid or NiMH battery
Honda EV+	455 per month			NiMH
Nissan Hypermini		23 350 with subsidy	36 000	Li-Ion
Peugeot 106 Elect.		14 700 excluding battery	27 000	NiCd
Renault Clio Elect.		16 000 with subsidy	27 400	NiCd
Solectria Force		34 000		Lead-acid
Solectria Force		67 500		NiMH
Toyota RAV4	499 per month	45 000		NiMH

Table 12.2   Powertrain/driveline cost comparisons (from Reference 1)

Vehicle	Fuel	Cost	Fuel savings (%)	Emissions*
Conventional IC	Gasoline	100	0	ULEV/SULEV
Conventional IC	Diesel	100	30	ULEV/SULEV(1)
Conventional IC	CNG/LPG	105–110	0	EZEV
Conventional IC	Hydrogen	200	0	ZEV
Hybrid IC	Gasoline/diesel	130–135	46–65(2)	ULEV/SULEV(3)
EV range extender	Gasoline/diesel	170–180	35–60(2)	ZEV(4)
EV pure battery	Electricity	150–200	60–80	ZEV
Fuel-cell	Compressed hydrogen	1 000+(5)	?	ZEV
Fuel-cell	Liquid fuel	?	?	?

(1) Possible problems with NO_x/particulates   * California Emission Standards
(2) Urban use   ZEV – Zero-emission vehicle
(3) Can have electric range for urban areas   ULEV – Ultra-low EV
(4) Qualifies for high ZEV credits   SULEV – Super ultra-low EV
(5) Cost will reduce in mass production   EZEV – Equivalent zero EV

only mode, while still being able to travel long distances and recharge the battery using the heat engine when being driven outside an urban area. It should be noted that this is not possible with the currently available production hybrids, since they do not have the ability to operate for more than very short distances in 'electric-only' mode.

12.2 Cost of batteries and fuel-cells

As has already been said, the cost of batteries and the other available alternative power storage devices is the most critical factor in determining the ability of electric vehicles to compete on a value-for-money basis with conventional internal combustion-engined vehicles. This is because not only purchase cost but also a major part of the running costs is determined by the cost of the battery and the frequency of replacement necessary. Firm figures for battery costs are difficult to obtain but best estimates of probable large-scale production costs are given in Chapter 5, Table 5.1. If we compare the estimated large-scale production cost of the currently most widely used advanced batteries (nickel-cadmium and nickel-metal hydride) at $250–$300 per kWh, with the cost of lead-acid at $60 per kWh, it is seen that the advanced batteries cost at least four times as much. However, these advanced batteries have at least twice the operating life, this being shown in Table 5.1 by the number of deep discharge cycles possible before battery replacement is necessary.

Taking a typical electric car that requires a battery capable of storing 25 kWh of energy, the initial cost of a lead-acid battery would be under

$2 000 compared to nearly $8 000 for an advanced battery. However, the lead-acid battery would need to be replaced every three years, while the NiCd and NiMH batteries would continue in use for at least twice as long so that advanced battery replacement costs spread over the life of the vehicle would only be twice those for the lead-acid battery. At present, advanced batteries are only produced in small quantities and under these circumstances the price may be as high as ten times that of lead-acid batteries.

The hope for the longer term is that advanced battery prices will come down to not more than twice that of lead-acid batteries while retaining their present operating life. The best hope for achieving this appears to be with the lithium-ion battery since this has potentially a much simpler manufacturing process than its competitors.

The major alternative to chemical batteries as a power source in electric vehicles is the fuel-cell. This has the great advantage, when fuelled directly by hydrogen, of retaining the zero-emission capability of the pure electric car. Alternatively, when fuelled by hydrogen generated from a reformer that processes a hydrocarbon or alcohol primary fuel (see Chapter 11), emissions should be low enough to meet the ultra-low emission category. Hydrocarbon or alcohol fuels can be stored in a tank on-board that is refilled in the same way as the gasoline tank in a conventional vehicle. It has been suggested in [1] that using fuel-cells increases driveline cost over conventional vehicles by over ten times even when they are directly fuelled with hydrogen, this implies even higher cost for a hydrocarbon- or alcohol-fuelled reformer/fuel-cell system. Currently fuel-cells are handmade and thus their high cost is not surprising. The critical test is how much fuel-cell and reformer costs can be reduced by large-scale production. It seems unlikely that the answer to this question will be known for at least ten years or that acceptably priced fuel-cell cars will be available much before 2020.

12.3 Hybrid costs

In the case of the heat engine/electric hybrid car cost competitiveness will be determined by the cost of the dual propulsion systems. The heat engine will normally be a conventional gasoline- or diesel-powered internal combustion engine with possible special lightweight components to help maximise fuel economy, its cost will be similar or slightly greater than a conventional vehicle powertrain system. The electrical propulsion system, which is almost all additional cost, consists of the battery, controller (to provide optimised control of both electrical and mechanical systems), propulsion motor or motors and the gearbox required to integrate the two drive systems. The battery would usually be expected to be substantially smaller than is required for a pure electric vehicle, its size being dependent on the distance over which the vehicle is required to run on electric power only. For example, the Honda Insight hybrid has a NiMH battery of only 0.94 kWh, all that is required to fulfil the 'motor assist' function for which the electrical propulsion system was designed. The Toyota Prius uses a 1.9 kWh NiMH battery that gives it the capability of travelling a short distance under electrical power only. Neither of these production hybrids are capable of operation for sufficient distance under electrical power alone to meet the Californian hybrid zero-emission requirements in a city such as Los Angeles, the function of the hybrid powertrain in the Honda and Toyota vehicles is purely to improve fuel consumption. This arrangement may, however, meet the objective of qualifying for the 'ultra low' or 'super ultra low' emissions categories of the Californian regulations.

The hybrid system can never therefore be equal in cost to the single drive system of the conventional car. What has to be established is whether the improvement in fuel consumption and consequent reduction in emissions is worth the additional cost. Future developments in internal combustion engines and their controls are expected to produce significant further reductions in their fuel consumption and this will continue to make it difficult to justify the additional cost and complexity of the hybrid.

12.4 Electricity supply and charging

Electric cars in which stored electricity is the only method of propulsion require recharging from a fixed supply at frequent intervals. Based on an average energy consumption of 0.2 kWh per km and a best battery charge/discharge efficiency of 80 per cent, an electric car which travels a typical commuting distance of 40 km (25 miles) per day will require a minimum of 10 kWh of energy to fully recharge its batteries overnight. Assuming an eight-hour charging period at constant current this would require an average of 1.25 kW from the supply mains over the whole period.

The cost of electrical energy varies around the world with Australia and New Zealand having some of the lowest prices. It is difficult to determine current real costs because of the variations within individual countries, but

Table 12.3   Estimated night-charging rates (1999)

Country	Industrial rates comparison (1998) (US cents/kWh)	Estimated night-charging rate (US cents/kWh)
Australia	4.5	3.5
France	5.6	4.3
UK	6.2	4.7(1)
USA	7.1	5.5
Germany	7.3	5.6
Italy	8.7	6.7
Japan	16.9	13.0

(1) Actual night rate (1999)

Table 12.3 gives approximate current comparative industrial rates for various countries and also shows estimated night rates that could be expected to be applicable to electric cars being charged overnight. Taking an average night-charging rate of 5.5 US cents/kWh, the cost of recharging for our 40 km per day commuter would therefore be $0.55 each night. Assuming comparable use of the car at weekends to the weekday commuting use this equates to about $200 per year. To this must be added the replacement cost of the battery which, for a production lead-acid battery costing $2 000, would be about $700 per year if we assume a three-year life, or for a production advanced battery costing $8 000 about $1 400 per year assuming a six-year life. Effectively, therefore, combined fuel and battery costs would total about $900 per year for the lead-acid battery car and $1 600 per year for the advanced battery car. This compares to fuel costs per year for a conventional small gasoline-fuelled car covering a similar distance each year of about $3 000 in Europe and $1 000 in the USA.

12.5 Charging at home and away

In Section 12.4 charging costs were estimated based on the low night-time rates at which domestic power is currently supplied in many countries. If large numbers of electric cars were being charged overnight the power companies might decide that this discount for load-levelling overnight loads should no longer apply. This could double the cost of overnight charging and must be factored into running cost calculations when electric vehicle use becomes widespread. Charging when away from home is also likely to be much more expensive. Car park charging points will be relatively expensive to install and maintain and this overhead cost will be loaded onto the price per kWh used. These installation and maintenance costs will be even greater for the fast-charging stations expected to be required on the inter-city roads since they will require large power installations of the order of 1 MW each (see Chapter 7). They also may require costly electrical power storage systems. These issues all remain to be clarified and resolved but they do indicate that electricity costs for electric cars are not likely to be as low in relation to battery costs as has sometimes been assumed. They also show that provision of adequate charging facilities will constitute a major change to the way the electrical infrastructure is organised and may offer considerable opportunities to those prepared to invest in them.

12.6 Can the electric car compete economically?

By 2025 the pure electric car with either a low-cost lead-acid battery, or medium-cost mass-produced advanced battery (probably lithium-ion), should be fully competitive on cost when made in production numbers with the conventionally powered small car of that time. In fact it seems likely it may be somewhat cheaper because of the relative simplicity of a stand-alone electric drive and the likely reductions to be expected by 2025 in power inverter and control system costs. The relative costs between conventional and electric drive will also be affected by the expected increase in cost of the internal combustion engines of that period. Whether the public will accept the inevitable range limitations will depend very much on the transport usage patterns then prevailing. Clearly emphasis is going to continue to be placed on reducing conventional car usage both to reduce emissions and to improve congestion. Hydrocarbon fuel prices will undoubtedly continue to increase faster than inflation, making electric power more economically attractive, and more people who have second cars are likely to see the electric option as being viable for them. The picture is slightly less optimistic for fuel-cell and hybrid electric cars, since generally they will still rely on hydrocarbon/alcohol fuels. However since they will offer excellent fuel economy and unlimited range, they should still have a market in direct competition with a new generation of small-engined conventional cars if their costs can be made competitive. In Chapter 13 under the heading 'The electric car of 2025' I have detailed the electric cars that I consider will be on the market in that year. It is my considered opinion that they will be sold in substantial numbers by that date and will lead the world into an era in which the electric vehicle will be in widespread use in most countries.

References

1   WEST, J.: 'Propulsion systems for hybrid electric vehicles', Colloquium on Electrical Machines Design for All-Electric and Hybrid Electric Vehicles, IEE, London, 28 October 1999

Chapter 13

Future developments

In this book I have tried to describe the current state of electric vehicle technology at the beginning of the year 2001. It seems clear to me that before the first 20 years of this century are over, increasing pressure on the environment will result in increasingly severe regulation of emissions from vehicles in urban areas. This, together with the rocketing cost of hydrocarbon fuels, will ensure that personal transport needs will have to be met by a mixture of highly economical, relatively low-performance gasoline- or diesel-fuelled cars and by a range of electric cars of various types.

Because of the high price of hydrocarbon fuels it will be necessary to use the fuel that is available in a much more efficient and environmentally friendly way than at present. One way is to burn coal, gas or heavy oil fuels in a power station with sophisticated emission controls to drive efficient turbines to produce electricity that can then be used to charge batteries in electric cars. The overall energy efficiency, taking into account all the energy losses measured from the extraction of the coal or oil from the ground through to the road wheels of a vehicle, is not much better in the case of the pure electric vehicle than it is in the case of the internal combustion-engined vehicle (about 19 per cent for the electric vehicle versus 18 per cent for the ICE vehicle). However, the emissions are potentially much more controllable at the power station than in each car on the road and are also not usually emitted in urban areas. It is also possible to use electricity derived from non-polluting sources such as hydroelectric or wind/wave generation, or perhaps more controversially from nuclear plants. If sources of this type can be used then electric vehicles provide almost pollution-free transport.

In this chapter I shall be looking at potential future developments in the various electric vehicle systems and what effect these developments can be expected to have on the complete vehicle.

13.1 Propulsion methods

The future developments in electric motors seem likely to be confined to the drive towards lower costs, lighter weight, smaller volume and higher efficiency of the existing motor technologies. It seems very unlikely that there will be any dramatic breakthrough in the next 20 years which will produce a completely new type of motor suitable for electric vehicles.

AC motors will continue to dominate, particularly as electronic control systems reduce in cost and become more integrated. There is still likely to be some place for separately excited DC motors with simple control systems in low-cost electric cars, but this will be a relatively small market compared to the widespread use of AC motors in the electric vehicles that will be in use on our roads in 20 years time.

Since motor weight and physical size is so important in an electric car the motor that is most likely to offer this must be the favourite for future use. At the present stage of development this looks likely to be the brushless DC motor, which in spite of its name is actually an AC synchronous machine with a DC/AC inverter (see Chapter 3). Perhaps the most important issue with this or any of the other AC motors is whether they can be made small, light and powerful enough to fit directly into the wheels of an electric car to make possible what is generally considered to be an ideal direct-drive propulsion system. Since at least two high-torque motors, probably in an axial rather than a radial field configuration and preferably with some gearing, would be required for an optimised in-wheel drive system, cost is also a vital issue. Improvements in the cost, magnetic performance, temperature rating and corrosion resistance of the permanent magnets used in these machines is probably the key to reaching this goal.

Induction (asynchronous) motors will undoubtedly also continue to be widely used for some time to come because of their simple construction, reasonable cost and low maintenance requirements, as will synchronous PM motors. It is uncertain whether developments in switched-reluctance machines will ever show the improvements in power output, weight, volume, reliability, torque variation and noise required to make it a serious contender for electric vehicle application. However, work at Newcastle University which is directed towards making it possible to use a conventional three-phase inverter drive with a switched-reluctance motor and which would reduce drive-system cost and give potentially higher output, may give this motor technology a lifeline.

13.2 Energy sources

This is the area where most changes and developments are to be expected in the next 20 years. In the last 20 years enormous strides have been made in the development of new battery technologies. These technologies are described in detail in Chapter 5 but in this section we will look particularly at their future development and how they may be used. The availability of batteries with twice (nickel-metal hydride) or more than three times (lithium-ion) the storage capacity of the familiar lead-acid batteries has opened up the possibility of cars with electric-only drive and an operating range of up to 400 km on one charge. So far the major inhibitor to the use of these advanced batteries has been cost, but this must surely be overcome within the next 20 years, particularly if potentially large numbers of sales are to be possible. Sadly, the high-temperature batteries (sodium-sulphur, sodium-nickel chloride and lithium-iron sulphide), in spite of their high storage capacities, appear likely to be a blind alley in electric car propulsion battery development. The complexity and cost involved in maintaining their high temperature (300–450 °C) when in use in an electric vehicle, combined with the corrosive or toxic materials they require to operate, seems likely to inhibit their further development and cost reduction.

Nickel-metal hydride (NiMH) batteries are being used widely to replace nickel-cadmium (NiCd) since they offer somewhat better performance without the toxicity problems of cadmium. An energy density of 70 Wh/kg (twice that of lead-acid) is available with a power density of 200 W/kg; this together with the ability to recharge the battery to 80 per cent capacity in 35 min makes the battery very attractive to electric vehicle manufacturers. A major disadvantage is a cost of about ten times that of lead-acid in experimental quantities and probably about four times in production quantities. The high cost is partially offset by an improved life of six to seven years, this compares to about three years for lead-acid.

In the longer term it seems probable that the lithium-ion battery will be the battery of choice. This is because of its high cell voltage (3.6 V), high-energy density of 100–125 Wh/kg (nearly twice that of NiMH), even better volumetric density (three times that of NiMH) and its low self-discharge, but also because of its potential for low-cost quantity production. This low cost is a consequence of the plastic lithium ion (PLI) technology that is employed to manufacture the cells. This involves coating a flexible laminate with a carbon powder to form the negative electrode, coating a second flexible laminate with a lithium manganese dioxide powder to form the positive electrode and sandwiching a polymer gel electrolyte between these two laminates. Because the laminates are flexible and can be rolled together as a continuous process, it is potentially possible to produce large quantities at relatively low cost. The laminate form also has the potential for making batteries of differing shapes that could fit more easily into vehicles in which space saving is at a premium.

Also in the longer term the metal-air batteries may have a part to play. Zinc, aluminium, magnesium or lithium electrodes can be used, but only the first two are likely to produce practical batteries. The big advantage of these batteries is the very high-energy density of 200–220 Wh/kg, sufficient for an electric vehicle to travel more than 400 km on one charge. Because of their low power density (aluminium-air 30 W/kg; zinc-air 80–140 W/kg) they do, however, require an auxiliary battery/energy store of high power density to provide the high power output required for acceleration and hill climbing. In operation, recharging is done by replacing the zinc or aluminium electrode and in most systems the electrolyte as well at a specially equipped recharge station. Some ingenious schemes for facilitating this have been developed including the use of zinc cassettes and zinc pellets instead of solid-plate electrodes. Replacement times of about ten minutes are claimed and if these methods can be shown to be suitable for the general public to use in routine recharging, then metal-air systems may well have a place among the advanced batteries available for use in the production vehicles of 2025.

There has been considerable interest recently in the use of fuel-cells as the energy source for the electric car of the future. The interest is understandable since fuel-cells offer direct conversion of hydrogen or hydrocarbon fuel to electrical energy without significant emissions other than water and at an efficiency from hydrocarbon fuel to road wheels of at least 30 per cent, with potential for considerable further improvement. This compares to an average efficiency for an internal combustion engine and gearbox system of 20–25 per cent with little prospect for significant improvement in this efficiency because of the limitations of the Carnot cycle. The operation of the fuel-cell is described in detail in Chapter 6 but it can best be thought of as a device for allowing hydrogen to combine with oxygen from the air in a controlled way producing electrical energy with water as a by-product. Hydrogen can be stored to fuel the cell directly and future developments in this area using graphite nano-fibres as an absorbing medium may make it possible to store large quantities of hydrogen at low to moderate pressure in a relatively small volume. This technology could have an important role to play in making on-vehicle hydrogen storage safer and more practical.

Alternatively, hydrogen can be produced chemically in a 'reformer' which is carried on the vehicle and fed by a hydrocarbon fuel such as methanol or gasoline. Development has reduced the size of both the fuel-cell stack and the reformer so that electric car-makers can now consider housing the assembly beneath the floor of the vehicle giving room for four seats and reasonable luggage space. There are, however, fundamental problems of complexity caused firstly by the low voltage (0.7 V) available from each cell, requiring at least 350 cells in a stack to produce a reasonable system voltage of 250 V. This problem is made worse by the need to provide equal amounts of hydrogen to each cell and also to remove the water generated by the hydrogen/oxygen recombination process. Operation at very high or very low ambient temperatures also presents problems, particularly in respect of the risk of rupture of fuel-cell stack components under freezing conditions. System complexity is increased by the need to supplement the power provided from the fuel-cell by means of a high power density auxiliary battery to supply starting and acceleration power demand. The big advantage of fuel-cell technology is the ability of the fuel-cell vehicle to travel a similar distance to a conventional car before it is necessary to refill with hydrogen or hydrocarbon fuel. There is no reason why this refilling should take much longer than the time currently taken to fill the gasoline tank on a conventional car, so it should be possible to operate a fuel-cell electric car in exactly the same way as one using an internal combustion engine. There are, however, some safety issues in handling hydrogen in refuelling stations still to be resolved.

Considerable work is currently in progress by the major car manufacturers and specialist fuel-cell suppliers to develop prototype and experimental fuel-cell cars (see Table 11.1) but the earliest that any manufacturer is committing to having a production vehicle on the road is 2004. All the cars on which information is available use either direct storage of hydrogen or a methanol reformer, with no use to date of a gasoline reformer. Daimler-Chrysler are reported to be working on a demonstration version of a gasoline reformer system [1] and consideration is also being given by other researchers to fuel-cells which can operate directly on a mixture of methanol and water. The choice of high power density auxiliary battery/energy store is not known for most manufacturers but interestingly Mazda are using a supercapacitor in the Demio-FCEV and the Renault/Volvo European Project uses an NiMH battery in the Fever.

In my opinion the future for fuel-cell-powered vehicles is bound up with the resolution of the problem areas discussed above together with the development of gasoline reformers and/or of directly fed fuel-cells. The production of fuel-cell vehicles in any quantity will certainly require fuel-cell stacks and reformers of a size suitable for underfloor mounting at a cost which makes them competitive with production versions of advanced batteries such as lithium-ion. The only way to progress towards this cost target is to develop production-line methods to replace the hand-assembly used at present.

The only other method that has the potential to store sufficient energy to propel a car over what must be an eventual target distance of 400 km between charges is the flywheel. Flywheel technology has been described in detail in Chapter 5 and it was shown that large quantities of energy could be stored if the rotational speed of the flywheel can be increased to very high levels. It is not possible to do this using a conventional metal flywheel as the tensile strength of the material is insufficient. It has, however, been shown that by making a flywheel of carbon fibre it can be run up to at least 100 000 rpm with potential to increase this to 200 000 rpm without exceeding the tensile strength limits of the material. By the use of a number of small flywheels rotating at 200 000 rpm and mounted in contra-rotating pairs (to cancel gyroscopic effects), it has been suggested that it would be possible to store enough energy to obtain a vehicle range of 400 km.

Even assuming that this quantity of energy can be stored by this method at an economic cost, there are major safety issues of containment of the fly-wheel if it were to disintegrate in an accident, although this is potentially much easier with a carbon-fibre flywheel than with one made of solid metal. There is still a great deal of work to do to show that it is practicable to store large amounts of energy in a high-speed flywheel and that such flywheels can be manufactured and assembled on a production basis at a cost that is acceptable. If this proves to be possible, however, then the major advantage offered by flywheels of not requiring replacement during the life of the car, or losing storage efficiency with use, could make them a very attractive method of energy storage.

A number of organisations, principally in the USA, are working to produce flywheels suitable for both auxiliary power storage in hybrid vehicles and as the major power storage system in electric vehicles. However, all the indications are that it will be some years before a viable commercial product becomes available using this technology.

Short-term energy storage is also possible using supercapacitors, hydraulic accumulators or compressed air. Of these, the supercapacitor is probably the only method likely to be used to any significant extent. It is particularly suitable for use as the auxiliary energy store in a fuel-cell-equipped vehicle where it can currently provide power density levels of up to 1 kW/kg with potential for 4 kW/kg within three years. Mazda have used a supercapacitor in their Demio-FCEV fuel-cell vehicle as the auxiliary power source. There is good potential for production of these devices at an acceptable cost within the next five years.

13.3 Controls and power electronics

Future developments in this area beyond the current state of the art as described in Chapter 4 are likely to be mainly in increasing both the integration of the electronic control system and the range of functions it can control, as well as reducing its cost. Improvements in the cooling of those components carrying high currents will also permit a reduction in the physical size of the power electronics; more widespread use of liquid-cooling methods should help in this.

There is no indication that there are any dramatic developments to be expected in power switching devices in the near future. The MOSFET for peak voltages below 200 V and the insulated gate bipolar transistor (IGBT) for peak voltages up to 1 200 V will continue to be most widely used because of their superior control characteristics.

13.4 Charging

Charging of electric cars may be thought to be a straightforward matter of providing suitable charging points at places where electric cars are parked in urban areas or at dedicated charging stations set up on motorways and other trunk roads. Unfortunately, because of the long period required to charge an electric car at normal domestic charging rates it will be necessary to provide charging points in many of the parking places at supermarkets, stations and at the work place, if widespread use of electric cars is to become possible. This is the only way to ensure that users can 'boost charge' their vehicles when using them for commuting and shopping and avoid the embarrassment of some drivers coming to a halt in the street with a flat battery. It will also probably be necessary to provide a proportion of these charging points with fast-charge facilities for those drivers who have omitted to recharge their batteries overnight. There are also considerable potential problems on how the large number of drivers who park on the street overnight can have their cars connected by a vandal-proof connection to their domestic power supply. For this situation inductive charging pads in the road may provide a way of meeting this need. More accurate methods of measuring battery state of charge (SOC) will help drivers to feel confident of the range still available to them so that they can make intelligent decisions on when and where they need to recharge.

It might be thought that the availability of batteries such as lithium-ion, which can hold more than three times the charge of lead-acid batteries would ease this requirement for charging points in parking places. This would only be true if the same weight of batteries is used so that the full gain in range is obtained and the battery is maintained so that it is in a fully charged state at the start of the journey. In view of the limitations of the domestic power supply this may present some difficulties. A high-efficiency electric car such as the GM EV1 requires about 160 Wh/km (260 Wh/mile) on average to propel it under highway driving conditions [2]. Therefore to obtain a range of 400 km (250 miles) requires 64 kWh to be available from the battery. Because of the efficiency losses between charge and discharge which vary between 10 per cent and 30 per cent in most batteries, it will be necessary to provide up to 80 kWh to charge the battery from a 20 per cent minimum SOC to 100 per cent SOC. If constant-current charging can be used, this can be achieved in 8 h of overnight charging only if the domestic power supply is capable of providing an average of at least 10 kW over this period. As discussed in Chapter 7, because of the average power level per house ('diversity factor') of under 3 kW allowed by the power supply company, this would mean that the supply mains would require reinforcement if large numbers of people were charging their electric vehicles in the same area at the same time. In practice it seems likely that many electric-car designers might opt for using smaller advanced technology batteries which could be more easily housed and forego the increased range available, this would leave the requirement for a large number of boost charging points unchanged.

If electric cars are to be used for long-distance travel then fast-charge facilities will be required along the road network at which batteries may be recharged in the minimum possible period – 15 min for an 80 per cent charge. There are major infrastructure implications for the electrical supply system if these facilities are to be provided and these are considered in detail in Chapter 7 together with details of the various methods of charging. It is clear that it will take a number of years and a considerable investment of money for facilities on the scale required to be provided and this may turn out to be a greater inhibiting factor on the use of electric cars than any of the direct technical issues.

One way of overcoming these problems is to use cars powered by fuel-cells instead of chemical batteries or to go to hybrid vehicles. Fuel-cell and hybrid vehicles only require provision of adequate supplies of either hydrogen or suitable hydrocarbon or alcohol fuels. Hydrogen requires special handling facilities and is unlikely to be widely used in the shorter term; however, hydrocarbon or alcohol fuels such as gasoline and methanol can be handled by the existing fuelling infrastructure.

The use of batteries with much greater power storage capacity, such as lithium-ion, does hold out the possibility of a lower requirement for boost charging. However, unless this type of battery can also accept fast-charging – and whether this is possible is currently not clear although fast-charge times of one hour have been reported – it is unlikely to ease the infrastructure requirement significantly. Metal-air batteries also offer some relief from the charging problem but only if the electrode and electrolyte exchange that is required at recharge (see Chapter 5) is made rapid and straightforward.

Battery exchange has been suggested many times as a way out of the charging time difficulty and in the early days of motoring battery exchange stations existed in the USA to make possible 'touring' by electric car. In these days of complex vehicles and rigorous safety regulations, designing a vehicle in which a battery weighing up to 0.5 tonne can be easily and quickly exchanged and its high current connections remade with complete reliability has proved daunting and no vehicle using this technology is known to exist. The provision of battery exchange facilities also involves very substantial investment as discussed in Chapter 7.

Proposals have been made for continuous charging by non-contact methods from inductors buried in the road surface and even by means of radiated power by laser or microwave beams. Bearing in mind the high level of power transmission likely to be required to make these methods work, they are unlikely to be acceptable to a general public who already show nervousness over the very low levels of radiation from power lines and from cellular phone masts.

Because of the widespread infrastructure effects, the provision of adequate charging facilities for large numbers of electric cars is likely to be one of the most important factors in deciding if 'pure' electric cars are ever going to compete effectively with conventional cars. These problems do not exist for fuel-cell and hybrid electric cars, however, and it may be this factor rather than the improvements in battery technology (usually cited as the critical factor in making electric vehicles viable), which will decide which type of electric car we shall be using in 2025.

13.5 Vehicle design and safety

Almost all the design and safety issues in electric vehicles are bound up with battery or fuel-cell size, weight and operating voltage. Developments in advanced technology batteries such as lithium-ion will make possible electric vehicle propulsion systems comparable in size and weight to the internal combustion engine, gearbox and gasoline tank of a conventional vehicle, providing a restricted range for the electric vehicle is accepted. Operating voltages will, however, continue to be between 200 V and 350 V and therefore continue to require the special safety measures described in Chapter 8 to protect both the user and the service personnel from the risk of electric shock.

The reduction in vehicle weight expected from the use of advanced batteries will mean that it should no longer be necessary to reinforce structural components or strengthen suspension components, although there is clearly still a trade-off between range and battery weight and size. It seems probable that the electric cars of 2025 will be designed to be either lightweight with a restricted range of about 100 km for short distance urban use, or of greater weight with a range in the region of 400 km and the capability of fast recharge on long-distance trips.

Fuel-cell-powered vehicles will also gain from the development of more efficient fuel-cells operating either with small reformers using methanol or gasoline as the reformer feed stock or with fuel-cells specially developed to operate directly with hydrocarbon or alcohol fuels. Operation with on-board-stored hydrogen seems likely to continue to be inhibited by the safety issues particularly during refuelling. Complexity and difficulties with operation at low ambient temperatures still remain significant problems with this technology.

Vehicles using hybrid heat engine/electric systems will continue to provide full operational flexibility at a cost significantly above that of the vehicles using a single propulsion system, but by the use of sophisticated operating algorithms will give lower fuel consumption than conventional internal combustion-engined vehicles. However, developments in lean-burn internal combustion engine controls in which engine characteristics, such as valve-timing and inlet-manifold-tuning, can be optimised continuously are likely to make it more difficult to show a dramatic fuel consumption advantage for the hybrid. Where the hybrid is likely to score is in making it possible to have a vehicle that can operate with zero-emission in urban areas and which has long-range capability on the open road.

It is difficult to see what significant improvement can be made over the next 20 years to reduce the energy used in vehicle heating and air-conditioning systems other than by the use of better body insulation and heat-reflecting double glazing, both of which add further weight to the vehicle. It may always be the case that the electric vehicle user has to accept a lower standard of heating and cooling than is possible in conventional vehicles.

Other design and safety issues are described in Chapter 8. However, they seem unlikely to significantly affect the development of electric or hybrid vehicles over the next 20 years and will therefore not be discussed further.

13.6 Hybrid technology

Hybrid technology using the classic heat engine/battery-electric combination overcomes both the range and charging-time problems already discussed. However, the problem of the cost of having two propulsion systems remains, and no juggling of the relative sizes of the two systems and the control algorithms to be used is going to make the complete vehicle as cheap as a vehicle with only one propulsion system.

The two production hybrid cars currently on the market in the USA and Europe from Honda and Toyota (see Chapter 10) are being sold at about half their true production cost. It seems unlikely that this situation could improve sufficiently even with quantity production to make them truly competitive with either conventional internal combustion-engined cars or with battery-only electric cars using low-cost advanced batteries if and when these are eventually available. The main market for future hybrids seems likely to be with users who require a vehicle that can be run as a zero-emission electric only in urban areas but as an almost conventional vehicle on long trips. It should be noted that this is not possible with the two production hybrid cars mentioned above, as they have neither the battery capacity or the electric motor power capability for continuous electric only operation at normal vehicle speeds.

13.7 The electric car of 2025

Let us now try to bring together all the developments likely in the next 20 years with the electric car technology available now in the year 2001 and speculate on the types and features of the production electric cars that will be available to the general public by 2025.

My view is that there will be the following classes of electric car:

1. A lightweight all-electric car designed for urban use with either very low-cost, old technology batteries (probably lead-acid) or, if their low-cost potential has been realised, advanced lithium-ion batteries. It seems likely that a single AC induction or brushless DC drive motor operating through a fixed-ratio gearbox will be used to keep cost down. Low-cost power electronics and controllers with sophisticated control algorithms will be widely available by 2025 and will be used to maximise operating efficiency and energy regained from regenerative braking. The car will use all available waste heat combined with minimal electric heating to give a reasonable standard of comfort in cold weather and a high-efficiency low-power electric compressor pump for a restricted level of cooling in the summer. The vehicle body will have a high level of lightweight insulation and the glass will be double-glazed and infrared reflective to reduce to a minimum heat transfer between the external and internal air volume. Performance will be unexciting but adequate for use in urban areas and range will be a minimum of 100 km of a full charge. Charging will normally be overnight from the domestic power supply but the car will be equipped to accept a fast-charge where facilities are available. As discussed earlier in this chapter the success of urban electric cars is dependent on adequate facilities for charging in car parks being provided.

2. A more expensive longer range all-electric car using the best advanced batteries (probably lithium-ion). Developments in magnetic materials should make it possible to use high-power geared in-wheel propulsion motors using brushless DC motor technology. Controllers, as well as providing comprehensive energy management of all the energy-using and recharging equipment on the vehicle based on an electronic model of the vehicle held in memory, will also provide comprehensive driver information including true battery condition and miles to empty. Details of controllers with these facilities are given in Chapter 4. A higher standard of heating and air-conditioning will be provided but will be controlled by the energy management system to give prioritised shutdown if the battery SOC is low. Performance will be comparable to a gasoline-engined car with a range between charges of about 400 km. Charging will normally be overnight at a high rate (6–10 kW) from the domestic power supply but with the capability to accept 15–30 min of fast-charging at specially equipped highway stations at a rate of up to 150 kW. Cost is likely to be a major issue with an electric car of this type. My view is that a comparable selling price to that of a conventional car can be achieved, but only if low-cost high-capacity batteries can be developed. The best chance of meeting this requirement appears to be by using lithium-ion technology.

3. A fuel-cell-powered car using low-cost reliable fuel-cell stacks made on a production-line basis. Combined with a small auxiliary battery of high specific power it should be capable of providing a maximum output of

75 kW at about 250 V. It will operate either with a reformer using a hydrocarbon fuel, probably methanol or gasoline, or in the likely event of suitable catalysts being developed by 2025, with a hydrocarbon fuel fed directly to the fuel-cells. In-wheel geared propulsion motors of a suitable power are likely to be available and will be used in conjunction with a comprehensive energy controller which will maintain fuel-cell conversion efficiency at a maximum and optimise and prioritise energy usage in all vehicle functions. Sufficient waste heat will probably be available from the fuel-cell stack – which operates at 60–100 °C – to provide adequate heating in a car insulated in the way suggested above for the all-electric vehicle. Providing sufficient energy for air-conditioning presents less of a problem than with all-electric vehicles since the vehicle energy is stored in the tank of hydrocarbon fuel that is easily and quickly refilled. Performance will be comparable to a gasoline-engined car, and range is only limited by the capacity of the hydrocarbon fuel tank. Cost should be similar to that of a conventional car providing reliable production fuel-cell stacks can be produced and connection and fuel feed complexity overcome.

4. A hybrid car using an optimised combination of electric drive and heat engine. Optimisation will be dependent on the conditions under which the car is to be used. For example, if it is to be used mostly in an urban area in which zero-emissions are a legal requirement, then it will need to be capable of operation as an electric-only vehicle over sufficient distance to enable it to travel from one side of the area to the other. This precludes the arrangement where a small electric drive operates with a relatively large heat engine (a 'motor assist' system) to provide additional power during acceleration or hill climbing as in the Honda Insight and Toyota Prius. Hybrid systems in which the heat engine and electric drive are comparable in power, or in which the electric drive is used as the main source of power and the heat engine is of small capacity (a 'range extender' system), are required for this area of operation. Since whatever the configuration of heat engine/electric drive, the heat engine is normally capable of recharging the battery without any requirement for connection to an external electrical supply. This means that all the normal power hungry features of the conventional car can be provided with the possible exception of full heating and air-conditioning on long zero-emission cross-city trips using only electric drive. The problem with the hybrid remains the cost of having two propulsion systems, but for the user who requires the hybrid's capability of operating with very good fuel economy on the open road and with zero-emissions in the city this could be an acceptable solution.

I have no doubt that electric cars in all their differing forms will become increasingly important in the transport scene over the next 25 years. Environmental pressures and fuel costs will combine to ensure their increasing use, but this increasing use is critically dependent on the development of high-capacity batteries with a cost not much greater than twice that of lead-acid. Lithium-ion technology appears at present to offer the best hope of achieving this goal.

References

1　TRAN, D.: 'DaimlerChrysler fuel cell report'. Panel Presentation at NAEVI 1999, Atlanta, Georgia
2　'EV1 electric/specs/pricing', Internet address http:// www.gmev.com/specs

Index